A WORLD ENVIRONMENT ORGANIZATION

Global Environmental Governance Series

Series Editors: John J. Kirton and Konrad von Moltke

Global Environmental Governance addresses the new generation of twenty-first century environmental problems and the challenges they pose for management and governance at the local, national, and global levels. Centred on the relationships among environmental change, economic forces, and political governance, the series explores the role of international institutions and instruments, national and sub-federal governments, private sector firms, scientists, and civil society, and provides a comprehensive body of progressive analyses on one of the world's most contentious international issues.

Also in the series

Hard Choices, Soft Law:
Voluntary Standards in Global Trade, Environment and
Social Governance
Edited by John J. Kirton and Michael J. Trebilcock
ISBN 0 7546 0966 9

The Politics of Irrigation Reform:
Contested Policy Formulation and Implementation in Asia,
Africa and Latin America
Edited by Peter P. Mollinga and Alex Bolding
ISBN 0 7546 3515 5

Governing Global Biodiversity:
The Evolution and Implementation
of the Convention on Biological Diversity
Edited by Philippe G. Le Prestre
ISBN 0 7546 1744 0

Agricultural Policy Reform:
Politics and Process in the EU and US in the 1990s
Wayne Moyer and Tim Josling
ISBN 0 7546 3050 1

A World Environment Organization

Solution or Threat for Effective International Environmental Governance?

Edited by
FRANK BIERMANN and STEFFEN BAUER

Routledge
Taylor & Francis Group

LONDON AND NEW YORK

First published 2005 by Ashgate Publishing

2 Park Square, Milton Park, Abingdon, Oxfordshire OX14 4RN
711 Third Avenue, New York, NY 10017

Routledge is an imprint of the Taylor & Francis Group, an informa business

First issued in paperback 2018

British Library Cataloguing in Publication Data
A world environment organization : solution or threat for
    effective international environmental governance?. -
    (Global environmental governance)
    1. Environmental policy - International co-operation
    2. Environmental responsibility - International co-operation
    I. Biermann, Frank, 1967- II. Bauer, Steffen
    363.7'0526

Library of Congress Cataloging-in-Publication Data
Biermann, Frank.
    A World Environment Organization : solution or threat for effective international environmental governance? / by Frank Biermann and Steffen Bauer.
        p. cm. -- (Global environmental governance series)
    Includes index.
    ISBN 0-7546-3765-4
    1. Environmental responsibility--International cooperation. 2. Environmental policy--International cooperation. 3. World Environment Organization. I. Bauer, Steffen. II. Title. III. Series.

    GE195.7.B43 2005
    363.7'0526--dc22

                                                                    2004020205

ISBN 978-0-7546-3765-3 (hbk)
ISBN 978-1-138-37881-0 (pbk)

Contents

Part III: The Case Against a World Environment Organization

About the Authors

Steffen Bauer is the coordinator of the MANUS research group ('Managers of Global Change: Effectiveness and Learning of International Organizations') within the Global Governance Project (GLOGOV.ORG), and a doctoral candidate in political science at the Freie Universität Berlin.

Frank Biermann is professor of political science and environmental policy sciences at the Vrije Universiteit in Amsterdam, The Netherlands. He also heads the Department of Environmental Policy Analysis at the university's Institute for Environmental Studies (IVM), and is the director of the Global Governance Project (GLOGOV.ORG).

Steve Charnovitz is associate professor of law at George Washington University Law School in Washington, DC. Prior to joining the George Washington University faculty, he practised law at Wilmer Cutler Pickering Hale and Dorr. He was also a co-founder and director of the Global Environment & Trade Study (GETS).

Lorraine Elliott is reader in international relations in the Department of Politics and International Studies at the University of Warwick in Coventry, and programme director for the department's MA in International Relations. She also holds a fellowship in the Department of International Relations at the Australian National University.

Thomas Gehring is professor of international relations at the Otto Friedrich Universität Bamberg. His recent research and publications have focused on the operation and interaction of international institutions, and on governance within the European Union. He has also published widely on international environmental regimes.

Joyeeta Gupta is UNESCO-IHE professor of policy and law on water resources and the environment at the UNESCO-IHE Institute for Water

Education, and manager of the programme unit 'International Environmental Governance' at the Institute for Environmental Studies (IVM) at the Vrije Universiteit in Amsterdam.

John J. Kirton is associate professor of political science, a fellow of Trinity College as well as director of the G8 research group, and a research associate of the Centre for International Studies at the University of Toronto. He is also editor of the Ashgate series on *Global Environmental Governance*.

Adil Najam is associate professor of international negotiation and diplomacy at the Fletcher School of Law and Diplomacy at Tufts University. His recent research and publications have addressed international environmental policy with a particular focus on developing countries and issues of human development and human security.

Sebastian Oberthür is assistant professor at the Otto Friedrich Universität Bamberg. His recent research has focused on interaction between international institutions and on options for reforming international environmental governance. He has also published widely on international cooperation on climate change.

Konrad von Moltke is senior fellow at the International Institute for Sustainable Development in Winnipeg, Canada. His recent work has focused on international environmental policy in relation to international economic policy. He is also editor of the Ashgate series on *Global Environmental Governance*.

Foreword

We are living in a dynamic, ever changing world where scientific knowledge is expanding and the challenges of managing the planet for the benefit of all peoples and all life forms are becoming more acute with every day. It is a world, both practically and philosophically, very different to the one in which the United Nations Environment Programme (UNEP) was born more than thirty years ago at the United Nations Conference on the Human Environment in Stockholm, Sweden.

This is reflected in the last big global summit in Johannesburg, South Africa. Here delegates gathered not for a conference on the Human Environment but for a World Summit on Sustainable Development (WSSD). The environment is now firmly embedded in this new and evolving approach. It is no longer viewed in isolation. It is one of the three pillars that underpin our quest towards a world that balances the need for development with the conservation of the life support systems that are the foundation of our very existence, prosperity and joy.

More than ever society recognizes that to deliver a new and more just world will require a fairer and more equitable sharing of its riches. Currently, a small fraction of its citizens is enjoying unprecedented levels of wealth at the expense of the majority. The costs of these life-styles are being 'externalized' to the poorer parts of the globe. They are footing the bill whether it be in relation to the impacts of climate change, the result of mainly industrialized countries emissions, or in a loss of their natural resources, such as fish, to the richer consumers and more powerful, better equipped fleets, of the North. So the world's consumption and production patterns are now at the heart of a hot debate as are the economic structures that are driving globalization and trade.

Part of this debate centres on how to reform and evolve the global institutions responsible for delivering sustainable development so they can effectively and efficiently meet their responsibilities. One of these is UNEP. Steps have already been taken to strengthen the organization under the banner of International Environment Governance. Decisions and resolutions were adopted at UNEP's 21st Governing Council in Cartagena, Columbia, and are now incorporated in the 2002 WSSD Plan of Implementation, giving UNEP fresh impetus and greater clarity in its responsibilities.

Indeed, it is gratifying to note that countries accepted the importance of putting the so-called multilateral environmental agreements covering everything from the protection of the ozone layer to trade in endangered species, on a par with the World Trade Organization.

Membership of the organization is again expanding and a new structure, called the 'indicative scale of contributions', for improving UNEP's financial position has been agreed in principle if not in final detail. However, the debate, which has involved not only governments but also civil society and academia, is far from over, with some countries convinced that a World Environment Organization is needed and others expressing an interest in a United Nations Environment Organization. Others are not convinced and believe a strengthened UNEP will suffice. In this respect, I welcome the actions of President Chirac of France, who has done more than any leader to catalyze and invigorate this important debate.

Until now, we have been without a solid, authoritative and comprehensive book, covering the pros and cons of the various streams of ideas. So I thank the editors and writers of *A World Environment Organization: Solution or Threat for Effective International Environmental Governance?*

I believe it will make an important contribution to informing a wider audience on where we need to go.

Klaus Töpfer
Executive Director
United Nations Environment Programme

Preface

The debate on creating a world environment organization started some thirty years ago, and no end is in sight. Many papers, articles and policy briefs have been written on this subject, both supporting and rejecting the idea of a new international agency for environmental policy, with no clear consensus emerging after three decades of debate. The public policy domain appears equally divided, with some governments strongly in favour, and others strongly opposing the creation of a new world environment organization.

Many expected the 2002 World Summit on Sustainable Development in Johannesburg to address these issues and to agree on a consensual roadmap for the institutional and organizational reform of global environmental governance. However, in the preparatory process leading up to Johannesburg it became clear that no major decisions would be taken. Governments represented in Johannesburg agreed neither on a timetable for initial discussions on a new world environment organization or any similar agency, nor on a clear programme of other reform options that would not include a new organization. The issue appears to have been postponed for the post-Johannesburg phase, and it is most likely that the debate on the reform of international environmental governance, including the creation of a world environment organization, will continue.

To date, this debate has been conducted in a scattered manner, in venues that range from public speeches, newspaper articles and conference proceedings to articles in peer-reviewed journals. These articles and texts are found in diverse communities within and outside academia, which makes it difficult for scholars, students or practitioners to study all reform proposals.

So far, there has been no comprehensive book, neither a single-authored volume nor an edited volume, to which the interested researcher or decision-maker could turn. We were thus happy to accept the invitation of Ashgate Publishers to compile an edited volume with some of the key reform proposals and their critics. We are glad to have succeeded in assembling a group of accomplished experts in this field, with three voices in favour and three voices against a world environment organization, and two contributions that provide the general empirical and analytical background within which this debate is grounded. While some chapters are entirely new, others draw on earlier versions that have been substantially revised to account for current developments, in particular the 2002 World Summit on Sustainable Development. We hope that this first book exclusively focused on the question of a world environment organization will help practitioners and academics alike to orient themselves in the discussion, to form their opinion, and hopefully to join and further enrich the debate.

Such an edited volume would not be possible without the support of many people, in addition to the contributors. We wish to thank for their support John Kirton and Konrad von Moltke, the general editors of Ashgate's series on *Global Environmental Governance*, who have also contributed a chapter each to this volume. We are also grateful to four anonymous reviewers and Kyla Tienhaara for reviewing the entire volume. Beatrice Alders of the Vrije Universiteit Amsterdam, Steffen Behrle of the Freie Universität Berlin and Marc Heinitz provided invaluable editorial assistance. This project would also not have been possible without the overall support of the Volkswagen Foundation in Hanover, Germany, for the MANUS research group of the Global Governance Project GLOGOV.ORG.

Frank Biermann, Amsterdam, and Steffen Bauer, Berlin

Chapter 1

The Debate on a World Environment Organization: An Introduction

Steffen Bauer and Frank Biermann*

1   Introduction

Many a book has been written about the state of the world environment. This book will not offer yet another assessment of the global environmental crisis and of the ineffective societal response.[1] Instead, this volume will focus on possible reform options for the emerging yet still insufficient system of international environmental governance. In particular, we will discuss one reform proposal that has been around for more than thirty years, but has now received fresh attention: the creation of a 'world environment organization' within the system of the United Nations. Would a world environment organization contribute to the solution of the global environmental crisis—or would it rather hinder progress because it would create new problems instead of solving existing ones, or because setting up a new agency would simply require too many resources with no clear benefit?

*   Many thanks to Aarti Gupta and Philipp Pattberg for valuable comments on earlier drafts of this chapter. Funding for this research by Volkswagen Foundation, Hanover, is gratefully acknowledged.
[1]   For an overall comprehensive assessment, see the United Nations' third Global Environment Outlook, GEO-3 (UNEP 2002a). See also Schellnhuber and Wenzel (1998).

In addressing this question, the book seeks to contribute to the larger debate on redesigning the current system of international, or global, environmental governance. The concept of global governance has been developed in recent years in light of the emergence of new actors and political mechanisms at the global level, and of the increasing acceptance of only limited sovereignty of states in times of global interdependence.[2] Global governance, as opposed to traditional interstate politics, has been most pronounced in trade, finance and economic policy, but also environmental policy (Meyer et al. 1997; Mitchell 2002; Young 1994). Yet despite recent achievements in institutionalizing international environmental cooperation, the overall governance architecture, impressive as it may be in some respects, still fails to manage and halt global ecological deterioration. Effective environmental governance hence requires significant improvements of the status quo.

Creating a new world environment organization is but one proposal to address these shortcomings. The debate on a new agency has been underway for more than thirty years. All contributions to this book draw on the findings of these three decades of academic and public discourse, and elaborate the issues further in light of new experiences as well as arguments by those with opposing views. In the next section we summarize the debates and developments upon which the contributions to this volume build.

2   The Evolution of a Debate: Does the World Need a WEO?

In the policy debate about how to improve global environmental governance, the idea of a strong specialized environmental agency under the auspices of the United Nations has seen three peaks in attention from policy-makers and scholars: An initial one in the early 1970s, around the 1972 UN Conference on the Human Environment; a second one in the mid-1990s, this time coinciding with the 1992 UN Conference on Environment

---

[2]   On global environmental governance, see for example Biermann (forthcoming) and Dingwerth and Pattberg (forthcoming). For discussions of sovereignty under conditions of increased interdependence see, among others, Litfin (1998) and Krasner (1999).

and Development; and a third one in the context of the 2002 World Summit on Sustainable Development.[3]

Early Proposals

While proposals to create global institutions for environmental politics date back to the late nineteenth century,[4] it was the US foreign policy strategist George F. Kennan who started the debate on organizational aspects of what later evolved into today's global environmental governance discourse. To our knowledge, Kennan's call for 'an organizational personality' in international environmental politics (Kennan 1970, 408) was the first of its kind. In his time, Kennan's vision of an 'International Environmental Agency' encompassed only 'a small group of advanced nations' rather than a universal caucus that would include 'a host of smaller and less developed countries which could contribute very little to the solution of the problems at hand' (Kennan 1970, 410).[5] Other authors contributing to broaden and specify the early debate included Abram Chayes (1972) and Lawrence David Levien (1972).[6]

The response of the international community to this early debate was to set up the United Nations Environment Programme (UNEP). UNEP was created by the United Nations General Assembly following a decision adopted at the 1972 Stockholm Conference on the Human Environment (UNGA 1972). It is not a specialized UN agency, such as the World Health Organization, but a subsidiary body of the General Assembly reporting through the Economic and Social Council. The administrative costs of UNEP's headquarters—the Environment Secretariat in Nairobi—are covered by the general UN budget; an additional small 'Environment Fund'

---

3   For expectations in the immediate run-up to the Johannesburg summit see the collection of essays by Brack and Hyvarinen (2002).

4   See Holdgate (1999), Caldwell (1984), McCormick (1989), Meyer et al. (1997).

5   Given the Cold War, it is interesting to note that Kennan's proposal explicitly included 'communist and non-communist [states] alike' (while the socialist countries at that time had boycotted the 1972 Stockholm Conference); he explicitly praised the Soviet Academy of Sciences. See also Kirton, this volume, who argues for a world environment organization led by the 'Group of 8'.

6   See also Charnovitz (2002, 325).

supported by voluntary government contributions serves to finance specific projects. Originally, governments wanted UNEP to evolve into an 'environmental conscience' within the United Nations system that would act as a catalyst triggering environmental projects in other bodies and help to coordinate UN environmental policies. UNEP's founding resolution of 1972 explicitly speaks of a 'small secretariat'. UNEP was—and continues to be—a long way from an international organization commensurable with other sectoral bodies, such as the International Labour Organization (Levien 1972, Charnovitz 1993). Nonetheless, the creation of the UNEP secretariat in 1973 fundamentally altered the context of the organizational debate in international environmental politics and effectively halted it at the time.

The Second Peak

The debate about a larger, more powerful agency for global environmental policy was revived in 1989. The Declaration of The Hague, initiated by the governments of The Netherlands, France and Norway, called for an authoritative international body on the atmosphere that was envisioned to include a provision for effective majority rule.[7] Although with merely 24 signatories not representative of the international community, the declaration effectively helped to trigger a second round of proposals for organized intergovernmental environmental regulation. It included contributions by Geoffrey Palmer (1992), who argued for strong organizational anchoring of international environmental law under UN auspices; Steve Charnovitz (1993), who proposed an international environmental organization to be modelled on the International Labour Organization; and C. Ford Runge (1994) and Daniel C. Esty (1994) who, concerned about the emergence of an ever stronger world trade regime,

---

[7] Declaration of The Hague, 11 March 1989, reprinted in 28 International Legal Materials 1308 (1989). The declaration was signed by Australia, Brazil, Canada, Côte d'Ivoire, Egypt, Federal Republic of Germany, France, Hungary, India, Indonesia, Italy, Japan, Jordan, Kenya, Malta, New Zealand, the Netherlands, Norway, Senegal, Sweden, Tunisia, Venezuela, and Zimbabwe.

argued for a world, or 'global', environmental organization. This debate was fuelled by continuing doubts regarding the effectiveness of UNEP. A 1997 report by the United Nations Office of Internal Oversight Services heaped heavy criticism on the management and the overall performance of UNEP (UN OIOS 1997). The report argued that UNEP lacked a clear role and that it was not clear to staff or stakeholders what that role should be. Instead, much time and energy had been spent in paring down programmes, which had reduced the time to do environmental work. The report also found that UNEP's secretariat lacked efficiency and effectiveness. In 1998, Klaus Töpfer, a former chair of the Commission on Sustainable Development, was appointed as UNEP's Executive Director, and a number of organizational reforms were undertaken (see Elliott, this volume).

This did not, however, end the debate on a world environment organization that could replace UNEP. In the late 1990s, representatives of the UN system themselves became active participants, and some high-profile international civil servants openly supported the creation of a new environmental agency, including the former head of the UN Development Programme, Gustave Speth, as well as the WTO directors Renato Ruggiero and his current successor, Supachai Panitchpakdi. The UN Secretary-General Kofi Annan (1997), in his comprehensive programme for renewing the United Nations, also addressed the environmental responsibilities of the UN. In particular he proposed to reform the UN Trusteeship Council in order to safeguard the global commons, taking up an idea that had first been launched by Maurice Strong in 1988.[8] Furthermore, Annan called on the UN General Assembly to set up a task force, led by Klaus Töpfer, to assess the environmental activities of the United Nations (see UNSG 1998). Following the report of this task force, an Environmental Management Group was created within the UN system, and it was decided that the UNEP Governing Council shall meet regularly at ministerial level. While the direction of this reform is to be welcomed, it remains to be seen whether this incrementalism in strengthening UNEP will deliver the necessary

---

[8] For a detailed discussion of trusteeship in international environmental law see Sand (2003, 2004).

results in the future, or whether more fundamental reforms are needed. Klaus Töpfer himself emphasizes the nexus of developmental and environmental concerns and is thus reluctant to call for a specialized agency that would focus exclusively on the environment. Instead, Töpfer appears to support the creation of a strong World Organization on Sustainable Development.[9]

In the meantime, a number of governments have also come forward with semi-official initiatives for establishing a new global agency. At the 1997 Special Session of the UN General Assembly on environment and development ('Rio+5'), Brazil, Germany, Singapore and South Africa submitted a joint proposal for a world environment organization. These countries suggested, in the words of Helmut Kohl, then Germany's chancellor and chief architect of this four-country initiative: 'Global environmental protection and sustainable development need a clearly-audible voice at the United Nations. Therefore, in the short-term ... it is important that cooperation among the various environmental organizations be significantly improved. In the medium-term this should lead to the creation of a global umbrella organization for environmental issues, with the United Nations Environment Programme as a major pillar' (Kohl 1997). A similar position evolved in France, exemplified by the speech of Dominique Voynet, then French environment minister, on 6 July 2000 before a subcommittee of the European Parliament. The minister stated that a new organization would need to build on UNEP. Furthermore, both the World Health Organization and the International Labour Organization seemed to function as role models for the French initiative, and the World Trade Organization is mentioned as a body to which an environmental agency should serve as a counterweight (Voynet 2000).

---

[9]   See for instance Töpfer's contribution to the Global Governance Speakers Series on 28 February 2003 [www.glogov.org/upload/public%20files/pdf/events/speakers/toepfer.-pdf, last visited on 23 July 2004].

The Current Debate

This renewed interest among some governments spurred further academic input to the discourse. Most scholars active in the debate so far published refined versions of their earlier arguments (Charnovitz 2002 and in this volume, Esty and Ivanova 2001 and 2002, Runge 2001). In many countries, increased attention to the question of a world environment organization emerged at the national level. In Germany, for example, supporters and opponents of a new organization have engaged in intensive debates in academic and public policy journals following a discussion paper by Biermann and Simonis (1998).[10] In 2001, the German Advisory Council on Global Change (2001) adopted the case for a world environment organization and advised the federal government to continue to work towards such a new agency.

The broadening of the debate in the late 1990s also resulted in a wide variety of new views about what a world environment organization should or should not do. Bharat Desai (2000) provided an extensive legal analysis and examined prospects for shaping a United Nations Environment Protection Organization that would report to a newly mandated UN Trusteeship Council. Richard G. Tarasofsky (2002) discussed how UNEP and its Global Ministerial Environment Forum could be substantially strengthened, although, notably without changing the programme's legal status or name. Peter M. Haas (2002) argued for a Global Environmental Organization that would mainly serve to centralize the collection and dissemination of environmental expertise. Lee A. Kimball (2002) discussed the international institutional conditions under which UNEP might be transformed into a 'W/GEO'. She also identified a number of developing country concerns by arguing that a well-designed environment agency would provide essential groundwork for 'development applications' rather than encumbering them.

---

[10] See in particular Biermann (2000, 2001, 2002); Gehring and Oberthür (2000); Oberthür (2001); Unmüssig (2001).

John Whalley and Ben Zissimos (2001, 2002) added an economic perspective to a debate that had thus far been sustained by diplomats, international lawyers and political scientists. They built their argument for a new agency on economic theory, in particular cooperation analysis and the Coase theorem (Coase 1960), and argued for a world environment organization that would create global markets in which environmental goods could be traded for non-environmental assets, notably money. The new global agency would provide the organizational, legal and financial arrangements required for deals between actors with an interest in the environmental behaviour of others, who would then receive financial offers in exchange for certain policies. The world environment organization would assist in monitoring the deals; provide insurance cover; identify possible partners for deals from governments to local landowners; and create package deals among all interested actors that would minimize free-riding incentives (Whalley and Zissimos 2001; for a critique of this approach, see Biermann 2001). Also, the economist Jeffrey A. Frankel argued that 'UNEP should be replaced from scratch' (cited in Charnovitz 2002, 327), and Joseph Stiglitz, 2001 Nobel laureate for economics, called at the World Social Forum in Mumbai in 2004 for a complete overhaul of the international institutions that address sustainable development.

Three Models of a World Environment Organization

In view of this plethora of well-intended suggestions—Magnus Lodefalk and John Whalley (2002) alone have reviewed no less than 17 recent proposals for a new environmental organization—significant overlap and also some confusion do not come as a surprise. Virtually all proposals for a world environment organization that have been circulated in recent years can be categorized in three ideal type models, which differ regarding the degree of change that is required.

First, the least radical proposals simply advise upgrading UNEP to a specialized UN agency with full-fledged organizational status. Proponents of this approach have referred to the World Health Organization or the International Labour Organization as suitable role models. In this model, other agencies operating in the environmental field would neither be integrated into the new agency nor otherwise disbanded. The new agency in this model is expected to improve the facilitation and coordination of norm-building and norm-implementation processes in comparison to UNEP. This strength would in particular derive from an enhanced mandate and better capabilities of the agency to build capacities in developing countries. This differs from UNEP's present 'catalytic' mandate that prevents the programme from engaging in project implementation in the field.[11] Furthermore, additional legal and political powers that could come with the status of a UN specialized agency could enable such a world environment organization to approve by qualified majority vote certain regulations which are then binding on all members. Its governing body could be a general assembly that could adopt drafts of legally binding treaties that have been negotiated by sub-committees under its auspices. Such powers could exceed those entrusted to the UNEP Governing Council, which has initiated intergovernmental negotiations on a number of issues, but cannot adopt legal instruments by itself.

A second group of proposals seeks to go further and advocates a more centralized, or streamlined, architecture. These authors make their case by challenging the substantive functional overlap between the multitude of institutions and organizations that have a say in international environmental policy. Indeed, harmonizing multilateral environmental agreements continues to be a priority concern for UNEP's Open-Ended Intergovernmental Group of Ministers or Their Representatives on International Environmental Governance (UNEP 2001b, see also UNU/IAS 2002, 12-15). Consequently, centralization advocates call for a streamlining approach that would integrate existing agencies and programmes into one all-encompassing world environment organization. They argue that prospective gains in efficiency and better coordination of international environmental policy could outweigh the risks that often accompany centralization. The integration of environmental regimes could loosely follow the model of the World Trade Organization, which has integrated diverse multilateral trade agreements under one umbrella. According to

---

[11] There is evidence, however, for a tendency within UNEP to expand the limits to its mandate in this respect already (personal communication with officers at UNEP headquarters, September and October 2003). Then, upgrading UNEP would merely be a logical consequence—a case of 'form follows function'.

some scholars, this integrative effort could even include established intergovernmental organizations,[12] although historic evidence suggests that this might go far beyond the politically feasible.

The third and most far-reaching model is that of a hierarchical intergovernmental organization on environmental issues that would be equipped with majority decision-making as well as enforcement powers vis-à-vis states that fail to comply with international agreements on the protection of global commons. Some observers see this as the only option to overcome the free-rider problem that has long plagued international environmental politics. The Hague Declaration of 1989 seemed to have veered in the direction of an environmental agency with sanctioning powers, and at the end of the 1980s, New Zealand had suggested establishing an 'Environment Protection Council', whose decisions would be binding.[13] Yet support for hierarchical models remains scarce. Apart from the European Union, the only example for a quasi-supranational body is the UN Security Council, which enjoys far-reaching powers under Chapter VII of the UN Charter. However, while the prospective benefits of an 'environmental security council' remain a part of the overall discourse, such an organization does not appear to be a realistic option in the next decades—and it is open to doubt whether it would be desirable at all.

The Positions of the Critics

There are also sceptical voices and outspoken critics of a new organization. The former executive secretary of the secretariat to the Convention on Biological Diversity, Calestous Juma, for example, raises the concern that advocates of a central authority are diverting attention from the more pressing problems of environmental governance, as well as failing to acknowledge that the centralization of institutional structures is an

---

[12] Daniel C. Esty (1996, 111), for instance, has suggested that UN specialized agencies such as the World Meteorological Organization may be merged into a new environmental agency.

[13] United Nations General Assembly, General Debate Settlement at the 44th Session, 2 October 1989, Statement of the Right Honourable Geoffrey Palmer, Prime Minister of New Zealand. See also Palmer (1992, at 278ff).

anachronistic paradigm (Juma 2000). Sebastian Oberthür and Thomas Gehring (Gehring and Oberthür 2000; Oberthür and Gehring, this volume; Oberthür 2001) support these concerns based on their reading of institutional theory. Konrad von Moltke (2001 and this volume) argues in favour of decentralized institutional clusters to deal with diverse sets of environmental issues rather than entrusting all problems to one central organization. Peter Newell (2001, 2002) has warned against the costly crafting of a world environment organization and supports instead multi-level governance approaches that are based on the principle of subsidiarity. Most recently, Adil Najam (2002, 2003, this volume) has joined the debate in arguing vehemently against a world environment organization. At the same time, however, Najam's work seems to indicate the increasing convergence of views, since his proposal of upgrading UNEP to a specialized agency does not appear to be far away from proposals of scholars who argue in favour of a world environment organization.

Not least, UNEP itself has contributed to the debate in an attempt to respond to the challenges that are seen to result, at least partially, from its delicate organizational status. It established in February 2001 an Open-ended Intergovernmental Group of Ministers or Their Representatives on International Environmental Governance to systematically assess the existing institutional weakness, identify future needs and consider feasible reform options. This process included consultations with academic experts at a workshop in Cambridge and with representatives from civil society groups at workshops in Nairobi (UNEP 2001a), as well as the involvement of the United Nations University (UNU/IAS 2002).[14] However, while this forum has touched on some important organizational issues, including the universal membership and improved financing of UNEP, it left open the question of whether UNEP should be upgraded to a world environment organization.

---

14 For more details see Elliott, this volume.

The Way Ahead

In sum, different phases which the idea of a world environmental organization has lived through indicate that there is no consensus in sight, and that the debate is not likely to disappear. The current view of governments—as summarized in 2002 by the then president of UNEP's Governing Council, David Anderson of Canada—appears to be that a new UN agency on environmental policy could be an option for consideration, but only in the longer term (see UNEP 2002b, para. 12). In this vein, the Malmö Declaration of the UNEP-initiated first Global Ministerial Environment Forum called upon the 2002 World Summit on Sustainable Development to 'review the requirements for a greatly strengthened institutional structure of international environmental governance based on an assessment of future needs for an institutional architecture that has the capacity to effectively address wide-ranging environmental threats in a globalizing world' (Malmö Declaration, para. 24).

Johannesburg, however, did not deliver. Hardly any aspects of institutional reform were addressed in a meaningful way. In retrospect it seems that institutional reform was an issue under continuous consideration in the years leading to Johannesburg, then essentially neglected at the summit, and now re-emerging again as an item of substantive debate.[15]

The French government has taken the lead again by circulating a proposal to transform UNEP into an 'Organisation spécialisée des Nations Unies pour l'environnement',[16] which follows up on earlier French initiatives to replace UNEP by an 'Organisation mondiale de l'environnement' or an 'impartial and indisputable global centre for the evaluation of

---

[15] See also Kanie and Haas (2004, 5) who suppose 'a time lag between renewed interest in the [WEO] proposal at the academic level and the more recent interest in the idea from a policy perspective'.

[16] Proposition française de transformer le Programme des Nations Unies pour l'environnement en une Organisation spécialisée des Nations Unies pour l'environnement, 12 Septembre 2003 [on file with authors].

our environment'.[17] The reaction of other member states of the European Union to this proposal remains to be seen. One larger European country, Germany, has explicitly stated its support for the French proposal (Trittin 2004).[18] Overall, however, it seems likely that France will neither encounter a clear majority of EU countries in favour nor a clear majority against its new initiative. As often, the European Union presents itself divided rather than united. Naturally, such divisions amplify at the global level where even less ambitious proposals, such as universal membership of the UNEP Governing Council, remain controversial.[19]

3   The Scope of the Book

It is within the context of the recent, more moderate reform proposals and their critics that the contributions to this volume are located. We have organized the book in three parts.

The State of Global Environmental Governance

Part I provides two analyses of the general state of global environmental governance as an informative backdrop to the entire reform debate. Lorraine Elliott offers a comprehensive account of the United Nations' activities in intergovernmental environmental politics. She focuses in particular on the decade after the 1992 UN Conference on Environment and

---

[17] See the speeches by Dominique Voynet (2000), the then French environment minister, and by French President Jacques Chirac (1998). See also the speech by the French minister for ecology and sustainable development, Roselyne Bachelot-Narquin, who recently reaffirmed the commitment of the French government to strive for the creation of a UNEO together with like-minded countries (Bachelot-Narquin 2004).

[18] The German Minister of the Environment, Jürgen Trittin (2002, 12), has repeatedly emphasized that '[t]he German government strongly favours transforming UNEP into a global environmental organisation. We need a strong global environmental institution that can stand up to the WTO, the FAO and transnational corporations'. See also Kohl (1997).

[19] At the eighth special session of the UNEP Governing Council, held in Jeju in March 2004, the notion of universal membership was supported by the European Union and Switzerland but met with strong objections from the US, G77/China, Japan as well as several other countries (see IISD 2004 for details).

Development in Rio de Janeiro. This 'Earth Summit' diversified the environmental governance structure, notably by creating the Commission on Sustainable Development, by the proliferation of substantive environmental conventions, and, through the 'Agenda 21', by elevating the concept of sustainable development to the UN's core paradigm for the 21st century. A critical assessment of the post-Rio developments, including the 1997 Special Session of the UN General Assembly and the Secretary-General's Task Force on the Environment and Human Settlements, is thus essential in evaluating the prospects of large-scale reform in global environmental governance.

In a complementary chapter, Joyeeta Gupta analyzes the implications of a changing global environmental governance structure for the South. She highlights the position of developing countries in international negotiations as key to determining eventual political outcomes, no matter which reform proposal is at stake. Gupta then describes a complex set of challenges that face an increasingly heterogeneous 'South'. She outlines nine scenarios for institutional reform, two of which reflect the idea of a strong world organization at the centre of global environmental governance—either as an integrated world organization for sustainable development or a sectorally focused world environment organization. Gupta then considers the pros and cons of each scenario and concludes that virtually all reform proposals, including the less ambitious ones, will ultimately work against the development interests of the global South.

The Case for a World Environment Organization

Part II then presents three chapters that support the case for a world environment organization. First, Steve Charnovitz develops his argument by highlighting lessons that may be drawn from the case of the World Trade Organization as a real-life point of reference. He then discusses the difficulties of global approaches with respect to subsidiarity in environmental decision-making, trade-offs between genuine environmental concerns and economically biased notions of sustainable development, as well as the need for a unified organizational approach despite the complexity of the global environment. While many reform proponents emphasize the need for greater coherence or enhanced transnational cooperation in world environmental politics, Charnovitz emphasizes the need for political competitiveness. Thus, a world environment organization as envisioned by Charnovitz would also be a competitor vis-à-vis other institutions of global governance.

Frank Biermann develops another proposal on what a world environment organization would look like. He suggests upgrading the United Nations Environment Programme (UNEP) to a specialized agency of the United Nations and outlines three core objectives that such an agency could help to achieve: a better coordination of international environmental governance, improved assistance for environmental policies in developing countries, and a strengthened institutional environment for the negotiation of new conventions and action programmes as well as for the implementation of existing ones. He then discusses some major issues and concerns in the current debate on a world environment organization, including whether this body should focus only on global issues or also on local issues, how a world environment organization could relate to the concept of sustainable development and to the interests of developing countries, and to what extent civil society could play a role within such a new agency.

John Kirton then discusses why the creation of a world environment organization *outside the UN system* would best meet the interests of major industrialized powers, namely the 'Group of 8'. He confirms the urgency to improve environmental governance yet questions the centrality of the United Nations in doing so. Kirton draws on regional regimes in North America and shows that economic and trade concerns can and should effectively be integrated with environmental protection. He argues that rich countries should lead the way in adjusting lessons learned at the regional level to global needs. In fusing his rationalist advocacy of 'Group of 8' leadership with normative considerations, Kirton highlights not only the capacities but indeed the responsibility of industrialized countries to move ahead in protecting the global environment.

The Case Against a World Environment Organization

Part III then includes three chapters that argue against the creation of a world environment organization. First, Konrad von Moltke rejects the idea of a world environment organization and calls upon critics to develop alternative proposals for the reform of global environmental governance. He then presents his own alternative proposal: to cluster the numerous international environmental agreements in order to tackle institutional overlap and fragmentation in international environmental policy-making. He argues that effective global environmental governance does not ask for centralizing environmental agreements in one high-level organization, but requires individual governments to champion well-designed clusters that address environmental macro-issues. In order to balance the diversity and complexity of the ecological crisis with the need for coherence and efficiency, von Moltke proposes separate clusters of multilateral environmental agreements, including the atmosphere, hazardous substances, the marine environment, and extractive resources.

Sebastian Oberthür and Thomas Gehring, two outspoken critics of proposals to create a world environment organization, base their arguments on their interpretation of institutionalist theory. They point to a number of organizational pathologies and reject key arguments of reform advocates, in particular those related to gains in efficiency and effectiveness. They argue that the political effort and resources required to create a world environment organization would spur a multitude of unintended side-effects, even as the desired long-term benefits remain uncertain and questionable. Ultimately, Oberthür and Gehring maintain that political attention and scarce resources should not be distracted for experiments with organizational reform as they could much better benefit the state of the global environment if deployed through the current institutional structure.

As a third critical contribution, Adil Najam argues that scholars who propose the creation of a world environment organization start on the wrong foot. His position is based on the premise that 'organizational tinkering' is little more than a superfluous distraction from the ultimate causes of the governance crisis plaguing international environmental politics. Among the more pressing issues, Najam highlights the demise of the Rio compact on sustainable development and a need for a society-centric view of environmental policy. One should not consider a new 'super-organization for the environment', he argues, but rather acknowledge the significant achievements of the UN Environment Programme and seek to further strengthen the latter's capacity through a number of feasible reform steps.

In the concluding chapter, the editors attempt to round up the debate for and against a world environment organization drawing on the various contributions in this volume. We discuss the key arguments, place them in the wider context of the global governance agenda, and try to identify some middle ground between the different approaches that address the prevalent crisis of environmental governance. We conclude by pointing to a number of areas where we believe that more research and debate is still needed.

This book is meant as a first comprehensive volume to present key arguments and criticisms in the debate on the organizational problems of global environmental governance and, in particular, on the potential role of a world environment organization. We hope that the book will make a meaningful contribution to inspire pragmatic debate on this central issue among scholars and decision-makers alike.

References

Annan, K. (1997), *Renewing the United Nations: A Programme for Reform*, UN Doc. A/51/950 of 14 July 1997, New York.

Bachelot-Narquin, R. (2004), 'Intervention de la ministre de l'écologie et de la développement durable', speech delivered at the conference 'Gouvernement internationale de l'environnement', Institute du développement durable et des relations internationales (IDDRI), Paris, 15–16 March [available at www.iddri.org/iddri/html/themes/archi/-gie.htm].

Biermann, F. (2000), 'The Case for a World Environment Organization', *Environment*, vol. 20 (9), 22–31.

Biermann, F. (2001), 'The Emerging Debate on the Need for a World Environment Organization: A Commentary', *Global Environmental Politics*, vol. 1 (1), 45–55.

Biermann, F. (2002), 'Strengthening Green Global Governance in a Disparate World Society: Would a World Environment Organization Benefit the South?', *International Environmental Agreements: Politics, Law and Economics*, vol. 2, 297–315.

Biermann, F. (2005), 'The Rationale for a World Environment Organization', in F. Biermann and S. Bauer (eds), *A World Environment Organization. Solution or Threat for Effective International Environmental Governance?*, Ashgate, Aldershot, 117–144.

Biermann, F. (forthcoming), 'Global Governance and the Environment', in M. Betsill, K. Hochstetler and D. Stevis (eds), *The Study of International Environmental Politics*, Palgrave Macmillan, Basingstoke, UK.

Biermann, F. and S. Bauer (2004), 'Assessing the Effectiveness of Intergovernmental Organizations in International Environmental Politics', *Global Environmental Change. Human and Policy Dimensions*, vol. 14 (2), 189–193.

Biermann, F. and S. Bauer (2005), 'Conclusion', in F. Biermann and S. Bauer (eds), *A World Environment Organization. Solution or Threat for Effective International Environmental Governance?*, Ashgate, Aldershot, 257–269.

Biermann, F. and S. Bauer (forthcoming), *Managers of Global Governance. Assessing and Explaining the Effectiveness of Intergovernmental Organisations*, Global Governance Working Paper, The Global Governance Project, Amsterdam, Berlin, Potsdam and Oldenburg [available at www.glogov.org].

Biermann, F. and U.E. Simonis (1998), *A World Organisation for Environment and Development: Functions, Opportunities, Issues*, Bonn, Development and Peace Foundation.

Brack, D. and J. Hyvarinen (2002) (eds), *Global Environmental Institutions. Perspectives on Reform*, Royal Institute of International Affairs, London.

Caldwell, L.K. (1984), *International Environmental Policy: Emergence and Dimensions*, Duke University Press, Durham, NC.

Charnovitz, S. (1993), 'The Environment vs. Trade Rules: Defogging the Debate', *Environmental Law*, vol. 23, 475–517.

Charnovitz, S. (2002), 'A World Environment Organization', *Columbia Journal of Environmental Law*, vol. 27 (2), 321–57.

Charnovitz, S. (2005), 'Toward a World Environment Organization: Reflections upon a Vital Debate', in F. Biermann and S. Bauer (eds), *A World Environment Organization. Solution or Threat for Effective International Environmental Governance?*, Ashgate, Aldershot, 87–115.

Chayes, A. (1972), 'International Institutions for the Environment', in J.L. Hargrove (ed.), *Law, Institutions and the Global Environment*, Oceana Publications, Dobbs Ferry, 1–26.

Chirac, J. (1998), *Speech by President Jacques Chirac at the Congress of the World Conservation Union (IUCN)*, 3 November 1998, Fontainebleau (on file with authors).

Coase, R. (1960), 'The Problem of Social Cost', *Journal of Law and Economics*, vol. 3, 1–44.

Desai, B.H. (2000), 'Revitalizing International Environmental Institutions: The UN Task Force Report and Beyond', *Indian Journal of International Law*, vol. 40 (3), 455–504.

Dingwerth, K. and P. Pattberg (forthcoming), *The Meaning of Global Governance. Towards Conceptual Clarity*, Global Governance Working Paper, The Global Governance Project, Amsterdam, Berlin, Potsdam and Oldenburg [available at www.glogov.org.].

Elliott, L. (2005), 'The United Nations' Record on Environmental Governance: An Assessment', in F. Biermann and S. Bauer (eds), *A World Environment Organization. Solution or Threat for Effective International Environmental Governance?*, Ashgate, Aldershot, 27–56.

Esty, D.C. (1994), 'The Case for a Global Environmental Organization', in P.B. Kenen (ed.), *Managing the World Economy: Fifty Years After Bretton Woods*, Institute for International Economics, Washington, DC, 287–309.

Esty, D.C. (1996), 'Stepping Up to the Global Environmental Challenge', *Fordham Environmental Law Journal*, vol. 8 (1), 103–13.

Esty, D.C. and Ivanova, M.H. (2001), *Making Environmental Efforts Work: The Case for a Global Environmental Organization* (= Working Paper 2/01), Yale Center for Environmental Law and Policy, New Haven (Conn.).

Esty, D.C. and Ivanova, M.H. (2002), 'Revitalising Global Environmental Governance: A New Function-Driven Approach', in D. Brack and J. Hyvarinen (eds), *Global Environmental Institutions. Perspectives on Reform*, Royal Institute of International Affairs, London, 5–18.

Gehring, T. and S. Oberthür (2000), 'Was bringt eine Weltumweltorganisation? Kooperationstheoretische Anmerkungen zur institutionellen Neuordnung der internationalen Umweltpolitik', *Zeitschrift für Internationale Beziehungen*, vol. 7 (1), 185–211.

German Advisory Council on Global Change (2001), *World in Transition: New Structures for Global Environmental Policy*, Earthscan, London.

Gupta, J. (2005), 'Global Environmental Governance: Challenges for the South from a Theoretical Perspective', in F. Biermann and S. Bauer (eds), *A World Environment Organization. Solution or Threat for Effective International Environmental Governance?*, Ashgate, Aldershot, 57–83.

Haas, P.M. (2002), *Science Policy for Multilateral Environmental Governance*, manuscript prepared for the United Nations University, University of Massachusetts, Amherst (Mass.) [on file with authors].

Holdgate, M. (1999), *The Green Web*, Earthscan, London.

IISD [International Institute for Sustainable Development] (2004), 'Summary of the Eigth Special Session of the United Nations Environment Programme's Governing Council/- Global Ministerial Environment Forum: 29–21 March 2004', *Earth Negotiations Bulletin*, vol. 16 (35), 2 April.

Juma, C. (2000), 'Stunting Green Progress', *Financial Times*, 6 July.

Kanie, N. and P.M. Haas (2004), 'Introduction', in N. Kanie and P.M. Haas (eds), *Emerging Forces in Environmental Governance*, United Nations University Press, Tokyo, 1–12.

Kennan, G.F. (1970), 'To Prevent a World Wasteland: A Proposal', *Foreign Affairs*, vol. 48 (3), 401–13.

Kimball, L.A. (2002), 'The Debate Over a World/Global Environment Organisation: A First Step Toward Improved International Institutional Arrangements for Environment and Development', in D. Brack and J. Hyvarinen (eds), *Global Environmental Institutions. Perspectives on Reform*, Royal Institute of International Affairs, London, 19–31.

Kirton, J. (2005), 'Generating Effective Global Environmental Governance: The North's Need for a WEO', in F. Biermann and S. Bauer (eds), *A World Environment Organization. Solution or Threat for Effective International Environmental Governance?*, Ashgate, Aldershot, 145–172.

Kohl, H. (1997), *Speech of Dr. Helmut Kohl, Chancellor of the Federal Republic of Germany, at the Special Session of the General Assembly of the United Nations*, 23 June, New York [on file with authors].

Krasner, S.D. (1999), *Sovereignty: Organized Hypocrisy*, Princeton, New Jersey.

Levien, D.L. (1972), 'Structural Model for a World Environment Organization: The ILO Experience', *George Washington Law Review*, vol. 40 (3), 464–495.

Litfin, K. (1998), *The Greening of Sovereignty in World Politics*, MIT Press, Cambridge (Mass.).

Lodefalk, M. and J. Whalley (2002), 'Reviewing Proposals for a World Environmental Organisation', *The World Economy*, vol. 25 (5), 601–17.

Malmö Declaration—GMEF/UNEP GCSS–6 (2000), *Malmö Ministerial Declaration*, Doc. UNEP/GCSS.VI/L.3, 31 May [available at www.unep.org/malmo/malmo_ministerial.-htm].

McCormick, J. (1989), *The Global Environmental Movement: Reclaiming Paradise*, Indiana University Press, Bloomington.

Meyer, J.W., D.J. Frank, A. Hironaka, E. Schofer and N.B. Tuma (1997), 'The Structuring of a World Environment Regime, 1870–1990', *International Organization*, vol. 51 (4), 623–51.

Mitchell, R. (2002), 'International Environment', in W. Carlsnaess, T. Risse, B. Simmons (eds), *Handbook of International Relations*, Sage, London, 500–16.

Najam, A. (2002), 'The Case against GEO, WEO, or Whatever-Else-EO', in D. Brack and J. Hyvarinen (eds), *Global Environmental Institutions. Perspectives on Reform*, Royal Institute of International Affairs, London, 32–43.

Najam, A. (2003), 'The Case against a New International Environmental Organization', *Global Governance*, vol. 9 (3), 367–84.

Najam, A. (2005), 'Neither Necessary, Nor Sufficient: Why Organizational Tinkering Won't Improve Environmental Governance', in F. Biermann and S. Bauer (eds), *A World Environment Organization. Solution or Threat for Effective International Environmental Governance?*, Ashgate, Aldershot, 235–256.

Newell, P. (2001), 'New Environmental Architectures and the Search for Effectiveness', *Global Environmental Politics*, vol. 1 (1), 35–44.

Newell, P. (2002), 'A World Environmental Organisation: The Wrong Solution to the Wrong Problem', *The World Economy*, vol. 25, 659–71.

Oberthür, S. (2001), 'Towards a Substantive Reform of Global Environmental Governance—Against the Creation of a World Environment Organization', in Heinrich Böll Foundation—Washington Office (ed.), *The Road to Earth Summit 2002*, Washington, DC, 57–64, [available at www.worldsummit2002.org/publications/roadtoes2002.htm].

Oberthür, S. and T. Gehring (2005), 'Reforming International Environmental Governance: An Institutional Perspective on Proposals to Establish a World Environment Organization', in F. Biermann and S. Bauer (eds), *A World Environment Organization. Solution or Threat for Effective International Environmental* Governance?, Ashgate, Aldershot, 205–234.

Palmer, G. (1992), 'New Ways to Make International Environmental Law', *American Journal of International Law*, vol. 86, 259–83.

Runge, C.F. (1994), *Freer Trade, Protected Environment*, Council on Foreign Relations, New York.

Runge, C.F. (2001), 'A Global Environment Organization (GEO) and the World Trading System', *Journal of World Trade*, vol. 35 (4), 399–426.

Sand, P.H. (2003), 'Trusteeship for Common Pool Resources? Zur Renaissance des Treuhandbegriffs im Umweltvölkerrecht', in S. von Schorlemer (ed.), *Praxishandbuch UNO. Die Vereinten Nationen im Lichte globaler Herausforderungen*, Springer, Berlin, 201–24.

Sand, P.H. (2004), 'Sovereignty Bounded: Public Trusteeship for Common Pool Resources?', *Global Environmental Politics*, vol. 4 (1), 47–71.

Schellnhuber, H.J. and V. Wenzel (1998), *Earth System Analysis. Integrating Science for Sustainability*, Springer, Heidelberg.

Siebenhüner, B. (2003), *International Organisations as Learning Agents in the Emerging System of Global Governance. A Conceptual Framework*, Global Governance Working Paper no. 8, The Global Governance Project, Amsterdam, Berlin, Potsdam and Oldenburg [available at www.glogov.org].

Tarasofsky, R.G. (2002), *International Environmental Governance: Strengthening UNEP*, UNU Working Paper [available at www.unu.edu/inter-linkages/docs/IEG/Tarasofsky.-pdf].

Trittin, J. (2002), 'The Role of the Nation State in International Environmental Policy: Speech by the German Federal Minister for the Environment, Nature Conservation and Nuclear Energy', in F. Biermann, R. Brohm and K. Dingwerth (eds), *Global Environmental Change and the Nation State: Proceedings of the 2001 Berlin Conference on the Human Dimensions of Global Environmental Change*, Potsdam Institute for Climate Impact Research, Potsdam, 10–13 [available at www.glogov.org].

Trittin, J. (2004), *Strengthening Environmental Protection in the United Nations*, speech delivered at the conference 'Gouvernement internationale de l'environnement', Institute du développement durable et des relations internationales (IDDRI), Paris, 15–16 March [available at www.iddri.org/iddri/html/themes/archi/gie.htm].

UN OIOS [United Nations Office of Internal Oversight Services] (1997). 'Report of the Office of Internal Oversight Services on the review of the United Nations Environment Programme and the administrative practices of its secretariat, including the United Nations Office at Nairobi. UN Doc. A/51/810 [available at http://www.un.org/Depts/oios/reports/a51810/-51-810e.htm, last visited 23 July 2004].

UNEP [United Nations Environment Programme] (2001a), *Reports of the Civil Society Consultations [Nairobi, 22–23 May] and Expert Consultations [Cambridge, UK, 28–29 May] on International Environmental Governance*, Doc. UNEP/IGM/2/2, 18 June [on file with authors].

UNEP (2001b), *Implementing the Clustering Strategy for Multilateral Environmental Agreements—A Framework*. Background paper for the fourth meeting of the Open-Ended Intergovernmental Group of Ministers or Their Representatives on International Environmental Governance, Montreal, 30 November–1 December 2001, UN Doc. UNEP/IGM/4/4 of 16 November 2001.

UNEP (2002a), *Global Environment Outlook—GEO–3*, Earthscan, London.

UNEP (2002b), *Report of the President of the United Nations Environment Governing Council for consideration by the Open-ended Intergovernmental Group of Ministers or their Representatives on International Environmental Governance*, Doc. UNEP/IGM/5/2, 16 January [on file with authors].

UNGA [United Nations General Assembly] (1972), *Resolution 2997 (XXVII): Institutional and Financial Arrangements for International Environmental Cooperation*, 15 December 1972, United Nations Organization, New York.

Unmüssig, B. (2001), 'Weltumweltorganisation: Quo vadis—Geisterdebatte oder neue Problemlösung?', *Rundbrief des Forums Umwelt und Entwicklung* 1/2001, 21–22.

UNSG [United Nations Secretary-General] (1998), *Environment and Human Settlements*. Report of the Secretary-General to the 53. Session of the United Nations General Assembly. UN Doc. A/53/463 of 6 October.

UNU/IAS [United Nations University/Institute of Advanced Studies] (2002), *International Environmental Governance. The Question of Reform: Key Issues and Proposals,* Preliminary Report (March 2002), United Nations University, Tokyo.

von Moltke, K. (2001), 'The Organization of the Impossible', *Global Environmental Politics,* vol. 1 (1), 23–28.

von Moltke, K. (2005), 'Clustering International Environmental Agreements as an Alternative to a World Environment Organization', in F. Biermann and S. Bauer (eds), *A World Environment Organization. Solution or Threat for Effective International Environmental Governance?*, Ashgate, Aldershot, 175–204.

Voynet, D. (2000), *Les priorités de la présidence francaise dans le domaine de l'environnement. Discours prononcé devant la commission on environnement-sante-consommation du parlement européen le 6 juillet 2000, à Strasbourg.* Speech, 6 July [on file with authors].

Whalley, J. and Zissimos, B. (2001), 'What Could a World Environmental Organization Do?', *Global Environmental Politics,* vol. 1 (1), 29–34.

Whalley, J. and Zissimos, B. (2002), 'Making Environmental Deals: The Economic Case for a World Environmental Organization', in D.C. Esty and M. Ivanova (eds), *Global Environmental Governance. Options and Opportunities*, Yale School of Forestry and Environmental Studies, New Haven, 163–80.

Young, O.R. (1994), *International Governance. Protecting the Environment in a Stateless Society*, Cornell University Press, Ithaca, NY.

Part I

Global Environmental Governance: Assessing the Need for Reform

Chapter 2

The United Nations' Record on Environmental Governance: An Assessment

Lorraine Elliott

1   Introduction

The World Summit on Sustainable Development (WSSD), held in Johannesburg in 2002, drew attention yet again to the challenges and perceived inadequacies of global environmental governance. The image of a crisis of governance is central to the global politics of the environment. The number of multilateral environmental agreements continues to grow (over 500 according to one estimate) (Wapner 2003, 6), but those agreements have often been less effective than is demanded by the environmental challenges they are intended to meet. They are often characterized by inadequate targets and commitments, permissive compliance and verification mechanisms and little in the way of effective sanctions.

Environmental institutions often lack formal competence or real power and are poorly funded. Despite what appears to be an increasingly sophisticated web of environmental principles—including sustainable development, the precautionary principle, the polluter pays principle, prior informed consent, intergenerational equity and common but differentiated responsibilities—those principles are still rarely implemented in any effective way. There has been little beyond a rhetorical commitment to important cross-sectoral concerns about new and additional resources, debt, inequitable trading relations or poverty alleviation, all of which featured in the 1987 report of the World Commission on Environment and

Development as crucial to the global pursuit of sustainable development. The overall state of the global environment continues to deteriorate, as the United Nations Environment Programme reported in its first Global Environment Outlook published in 1997 (UNEP 1997). UNEP repeated this message in its second global report, arguing that 'the global system of environmental management is moving ... much too slowly' (1999b, xxiii). UNEP's third Global Environment Outlook, published in 2002 just before the World Summit on Sustainable Development, was hardly any more comforting, observing that 'sustainable development remains largely theoretical for the majority of the world's population' (2002, xx). UNEP concluded, among other things, that 'the establishment of strong institutions for environmental governance is a prerequisite for almost all other policies' (2002, 396).

A number of institutional design models are proposed for responding to these challenges (see Vellinga et al. 2002; Charnovitz 2002; Lodefalk and Whalley 2002). One of the key issues in these discussions is whether a new or improved institution, such as a world environment organization, should be located within the United Nations. The reasons for the UN to play this central role seem obvious. The environmental problems facing the international community have a global reach—ecologically, economically and politically. This global reach is captured in metaphors that are intended also to reflect our shared vulnerability to and responsibility for global insecurities—'our common future' (WCED 1987), a 'global partnership' (UNCED 1992a), and 'our global neighbourhood' (Commission on Global Governance 1995). The UN is 'the only multilateral organization that is [almost] universal in its membership and global in its scope' (Strong 1991, 297). Although environmental protection is not mentioned in its Charter, the United Nations has been central to the development of a global environmental regime. It is difficult to think of an environmental issue with worldwide or global relevance (with the exception of the Antarctic or the World Commission on Dams) or an environmental agreement with universal membership, which is not in some way governed primarily under the auspices of some part of the United Nations.

However, there also is a growing feeling that the UN's capacity and legitimacy are in decline and that it is struggling to manage a more complicated global agenda. Indeed, calls for a world environment organization indicate a deep dissatisfaction with environmental practice under the UN (see, in particular, Kirton, this volume). As Newell observes, 'many proposals [for institutional reform] take as their starting point existing weaknesses in the way the UN addresses global environmental problems' (2002, 660). At worst, it is thought inefficient, ineffective and incapable of meeting the challenges of global change and the pursuit of a more secure world. At the very least, there is a widely held view that the 'UN system will have to make some vital changes if it is to remain relevant to the new multilateralism and to global governance' (Knight 1995, 244). The purpose of this chapter is not to argue for or against a world environment organization, nor for or against the United Nations as the preferred location should such an organization be established. Those issues are canvassed in depth elsewhere in this book. Rather, given that demands for a new institution arise because of perceived inadequacies in the present system of governance, this chapter examines what is thought to have gone wrong within the UN system as it has sought to meet the challenges of global environmental governance since the 1972 Stockholm Conference on the Human Environment.

A precise definition of global governance is elusive. In liberal-institutionalist terms, it is generally perceived as a more effective form of multilateral management. Institutional design is, however, only one component of the debate about improving global environmental governance and the assumption that more effective institutions will solve the problem are questioned (see, for example, Elliott 2004; Newell 2002; Gleeson and Low 2001). A more cosmopolitan or globalist approach emphasizes the importance of obligations that transcend borders, social justice and equity, and the relations of power and powerlessness, which condition the practices and outcomes of global governance (see, for example, Held 1996; Biermann 2002; Elliott 2002). In the context of these broad approaches, three dimensions of global governance are examined in this analysis of the United Nations: institutional competence, multilateral cooperation and participatory diplomacy; and normative reform.

Investigations of the UN's institutional competence in the field of global environmental governance, the first of these three dimensions, have focused on the need for a substantive and authoritative environmental body with the 'capacity to control and deploy ... resources' if it is to meet the objectives member states set for it (Commission on Global Governance 1995, 4). Improvements to institutional competence are also anticipated to require effective coordination to manage the growing institutional and policy complexity which characterizes the environmental agenda.

The second requirement of global environmental governance explored here is that the practices of environmental diplomacy must be cooperative and inclusive; in other words, governance must be democratic. The Report of the World Summit on Sustainable Development called for 'solid democratic institutions responsive to the needs of people' (United Nations 2002, 64). Cooperation among governments is required because states cannot unilaterally meet the challenge of global environmental change. At minimum, the role of the United Nations as an arena for negotiation and policy formulation anticipates that it will forge some kind of cooperative or collective sovereignty in situations such as environment degradation and sustainable development 'where [states] can no longer exercise it [sovereignty] effectively alone' (Strong 1973, x).

Inclusiveness is required on a range of democratic grounds. Agenda 21 called for 'democracy and transparency' as the cornerstones of environmental governance (UNCED 1992b, para. 38.2). Democratic pluralism acknowledges the legitimate interests of a range of non-state actors and stakeholders and the efficiency benefits that are met when they are involved in policy-making and standard setting. A cosmopolitan democracy ethic demands that governance responds to and includes those who are most often directly affected by environmental degradation but whose voices are rarely heard in the negotiating rooms—the poor, women, indigenous peoples.

Finally, as both the World Commission on Environment and Development and the Commission on Global Governance intimated, new norms are required to respond to global environmental challenges and enhance protection of the environment. UN Secretary-General Kofi Annan has exhorted governments to adopt a *new* ethic of conservation and stewardship (UNSG 2000, 63-5). Rule-making and standard-setting often feature more prominently in proposals for some kind of world environment organization than does norm-building. There are however general expectations that such a body would raise public consciousness and there is some attention to the kinds of principles that might be made central to any new body. The normative debate has tended to revolve around the issue of sovereignty and its relevance in a world now beset by the competing trends of globalization and fragmentation where the distinction between domestic jurisdiction and international relations is increasingly blurred (see, for example, Litfin 1998).

2   Institutional Competence and Reform

As noted above, authority and coordination are two key institutional demands of the UN and of any institution designed for global environmental governance. The UN has generally failed on the first count (although not necessarily for want of trying) and has struggled with the second. Institutional reform for the environment has become something of a permanent agenda item within the UN. The 1987 Brundtland Report called for a Board of Sustainable Development with powerful supervisory and coordination functions (WCED 1987, 318-19). The Hague Declaration of 1989, although not convened under UN auspices, recommended a new international authority with decision-making powers and the right to act in the absence of consensus to overcome the inadequacies of existing institutions. The 1992 United Nations Conference on Environment and Development was required to identify ways to improve the UN's role in dealing with the environment (UNGA 1989, para. 15(q)) and a full chapter in Agenda 21 (chapter 38) was given over to institutional issues.

Five years after UNCED, the Secretary-General's 1997 Programme for Reform gave priority (among other objectives) to strengthening the United Nations' capacity in its sustainable development and environmental dimensions. In 1998, as part of that reform process, the Secretary-General appointed a Task Force on Environment and Human Settlements, chaired by UNEP Executive Director Klaus Töpfer. Its report, which was submitted to the General Assembly later in 1998, reflected the view that 'the institutional fragmentation and loss of policy coherence ... had resulted in a loss of effectiveness in the work of the UN in the area of environment and human settlements' (Töpfer 1998, 1). The institutional problems in the UN were characterized as 'basic and pervasive' (UN Task Force 1999, para. 20). The Task Force made a number of recommendations that called for a results-oriented approach to effective inter-agency coordination and the strengthening of UNEP as the lead environmental agency within the UN. The reform issue was still on the table in Johannesburg. In much the same ways as at UNCED and UNGASS, WSSD was expected to 'address ways of

strengthening the institutional framework for sustainable development'
(UNGA 2000, para. 15(e)).

Nothing of the kind of institutional change anticipated in this
process of investigation and debate has yet surfaced. Despite the constant
focus on reform, responsibility for environmental protection and
sustainable development remains dispersed throughout the UN system.
Sandbrook describes it as 'Byzantine' (1999, 173). Over thirty specialized
agencies and programmes are now involved in the implementation of
Agenda 21 and the pursuit of sustainable development objectives (see
Blumenfeld 1994, 3). Only two UN bodies are *specifically* mandated as
environmental agencies or programmes. They are, of course, the United
Nations Environment Programme (UNEP) and the Commission on
Sustainable Development (CSD). Both arose from key UN global
conferences designed to encourage international efforts on environmental
degradation and sustainable development. Both have been compromised in
their form and function by political interests, and both have been plagued
by vague or ineffective mandates and persistent under-funding.

The United Nations Environment Programme, established in
accordance with recommendations from the 1972 Stockholm Conference,
was constrained from the outset by politics and turf battles. Developed
countries were reluctant to fund a new institution. Developing countries
were wary of any new UN agency which might seek to constrain
development. Other UN bodies were intent on jealously guarding their own
environment-related prerogatives and funding. It was for these reasons that
UNEP was established as a programme with a rather vague mandate—to
monitor, coordinate and catalyze.

Despite its lack of executive powers, UNEP has, in many ways, been
immensely successful in overcoming or at least working around its rather
limited mandate. This is a consequence, in part at least, of its competence
in establishing a strong environmental information and knowledge base,
and the political skills and legacy of active interventionist diplomacy
bequeathed by its early executive directors, particularly Mostafa Tolba. It
has coordinated international negotiations on ozone depletion, biodiversity
and desertification. It has forged partnerships with the World
Meteorological Organization (WMO) to advance the climate change debate
(including sponsorship of the Intergovernmental Panel on Climate Change)
and with the World Conservation Union (IUCN) to develop the World

Conservation Strategy. It has taken a lead role in regional seas protection, maintained the highly successful Global Environmental Monitoring System (GEMS) and responded to particular environmental crises such as the 'haze' incidents in Southeast Asia. It provides secretariat support for a number of international environmental agreements and is now responsible for producing the Global Environment Outlook reports.

UNEP has not, however, been able to overcome a general lack of political and financial support. Member states have provided only a 'meagre amount of support' (Wapner 2003, 8). For example, in the same year that governments committed themselves to the global partnership of Agenda 21 (1992), contributions to UNEP totalled only USD 62 million, leading to suggestions that core programme activities would have to be reduced to meet 1994-95 budget projections (Anonymous 1993, 119). UNEP's total resources in its first two decades totalled less than USD 1 billion. Its annual budget is less than that of most other UN agencies and some of the larger environmental NGOs (see French 1995, 29). Its professional staff is small (about 300 to 400) compared with its counterparts in national governments (see Biermann 2002, 298) although not when compared with other multilateral institutions such as the WTO or the IMF. These structural problems were compounded by management and political problems internal to UNEP. These included tension between senior management and staff, 'inefficient hiring practices [and] lack of transparency in decision-making' (Downie and Levy 2000, 362) and allegations of cronyism and problems of personal ambition (Sandbrook 1999, 172).

By the mid-1990s, there was a strong feeling that UNEP was facing an institutional crisis and that it was increasingly undemocratic and directionless despite its achievements to date (see Sandbrook 1999). At the 1997 General Assembly Special Session to review implementation of Agenda 21, the governments of Germany, Brazil, South Africa and Singapore submitted a joint declaration in which they called for UNEP to be 'reformed and strengthened ... [as] the world's environmental conscience' and for 'the establishment of a global environmental umbrella organization of the UN' (Joint Declaration 1997). As Charnovitz observes, the declaration 'did not meet with enthusiasm' (2002, 8). The only consensus that could be reached at UNGASS was to call *again* for an enhanced role and adequate funding for a revitalized UNEP, echoing a similar recommendation in

Agenda 21. Kofi Annan's 1997 reform agenda averred, once again, that high priority had to be given to UNEP to ensure the 'status, strength and access to resources it requires to function effectively as the environmental agency of the world community' (UNSG 1997c, para. 176). At UNEP's 20th Governing Council in February 1999 governments promised, as they had done many times before, to provide 'adequate, stable and predictable financial resources' (UNEP 1999a). They endorsed a new integrated work programme to support the 1997 Nairobi Declaration which proclaimed that UNEP should be the leading global environmental authority.

As noted above, the United Nations Conference on Environment and Development convened in 1992 with a sense that some kind of institutional renewal was required. The issue was contentious. Some participants favoured strengthening UNEP. Others wanted a new organization to provide a post-Rio focus and to demonstrate that delegates to UNCED really had done something. Agenda 21 recommended that UNEP should be better provided with financial resources but, French argues, other 'institutional decisions made at Rio undermined UNEP's prestige and confused its mission' (1995, 32). Chapter 38 of Agenda 21 recommended a new UN body, 'in a spirit of reform and revitalization of the United Nations system' (UNCED 1992b, para. 38.1). That body is the Commission on Sustainable Development (CSD).

CSD has two primary tasks. The first is to monitor and examine progress on the UNCED agenda, to 'illuminate' rather than to negotiate as a former Chair has put it (cited in IISD 1999). Its second role is to facilitate the integration of environment and development concerns within the UN system and among governments. In effect, this relieves UNEP of its own coordinating role but leaves it open to be 'coordinated' by CSD. CSD also has no executive, compliance or sanction powers. Within the limits of its original mandate the Commission has sought to adopt the UNEP line (although with less success) and expand its activities beyond its initial thematic focus on clusters of Agenda 21 issues. For example, CSD convenes a range of Multi-Stakeholder Dialogues on sustainable development issues. It was also responsible for establishing the Inter-Governmental Panel on Forests (later replaced by the Intergovernmental Forum on Forests and then by the UN Forum on Forests) and it acted as the preparatory committee for the UNGASS and WSSD, respectively the 1997 and 2002 reviews of Agenda 21.

Despite hopes that CSD would encourage member states to demonstrate their commitment to sustainable development, it is doubtful that it meets any of the reform expectations for a new substantive environmental body within the UN system. Nitin Desai, Under-Secretary-General for Policy Coordination and Sustainable Development, had argued that the Commission's success would depend on the 'political weight given to it by governments' (cited in Anonymous 1995, 163-4). That weight is not there. From the beginning, CSD has been weakened by compromise, consigned 'to the dungeon' as Imber puts it (1999, 330). Even today, it is not too unfair to accept Brenton's assessment that it is not much more than a 'UN talking shop with very limited impact on the world outside' (1994, 221).

Despite the urgency generated at UNCED, the Commission's first session in 1993 (CSD-1) was poorly resourced with 'only a handful of staff ... some ... borrowed from other departments' and 'difficulty on some days to secure enough rooms for the meetings' (Khor 1994, 103). This is hardly evidence of a strong institutional or normative commitment to either the Commission or Agenda 21. CSD sessions have been characterized by political disputes and a general reluctance to 'focus on concrete initiatives and proposals aimed at implementing the promises of Rio' (Blumenfeld 1994, 4). At CSD-4 in 1996, Chair Henrique Cavalcanti suggested in a frank admission to NGOs that he 'was not sure the CSD would be around after 1997' (Doran 1996, 100). The prediction was clearly premature but it is also clear that CSD has had difficulty in making itself heard within the UN and by member states.

At CSD-7 in 1999, Chair Simon Upton told participants that ministers had informed him that they would lose interest in the Commission if it failed to produce 'something substantive' (reported in IISD 1999), even though the ministers themselves were the key to successful negotiation and agreement. In an attempt to focus more on substantive outcomes, the Commission has moved to two-year 'implementation cycles' with review and policy sessions in alternate years. CSD-11 in 2003 adopted seven such cycles, effectively establishing a work programme through to 2017. Poverty eradication, production and consumption, and protection of the natural resource base remain the over-arching themes for each of these cycles, the first of which focuses more specifically on water, sanitation and human settlements.

The institutional reform agenda and concerns about institutional fragmentation were prominent in the prologue to the 2002 World Summit on Sustainable Development in Johannesburg. In response, governments 'created *three* new [institutions] to deal with the problem' (Charnovitz 2002, 18; emphasis in original). General Assembly resolution 53/242 established a Global Ministerial Environmental Forum (GMEF) tied to UNEP and requested the Secretary-General to strengthen UNEP by ensuring that it had adequate support and predictable financial resources (see UNGA 1999). At its first meeting in Malmö, Sweden in 2000, the Global Ministerial Environmental Forum stressed the importance of a 'greatly strengthened institutional structure for international environmental governance' and called for UNEP's role to be 'strengthened and its financial base broadened and made more predictable' (GMEF 2000, para. 24). Resolution 53/242 also approved the Secretary-General's proposal for an Environmental Management Group (EMG) to enhance inter-agency coordination in the field of environment and human settlements. UNEP's Governing Council, in Decision 21/21 of 2001, established an Open-ended Intergovernmental Group of Ministers or their Representatives to examine options for strengthening international environmental governance.

There was little new at WSSD either. Despite suggestions of a 'certain institutional effervescence' (Thomas 2000, 13) prior to the Summit, the language and recommendations adopted in Johannesburg were neither original nor inspiring. France and Germany led an unsuccessful push for a world environment organization (see Chirac 2002). The Plan of Implementation adopted at Johannesburg only endorsed, in a very general way, earlier decisions to strengthen UNEP and improve coordination with the UN system. It called for CSD to continue to 'serve as a forum for consideration of issues related to the integration' of various dimensions of sustainable development and argued that the Commission should be 'strengthened' (UN 2002, 67) although it gave little insight as to how this might happen.

3   A Permanent State of Reform

For many, the difficulties faced by UNEP and CSD are evidence that the UN has become unwieldy and unresponsive, characterized by demarcation and duplication of responsibilities which are poorly coordinated. The need for better coordination of environment and sustainable development objectives within the UN is driven also by the exigencies of geography. UNEP resides in Nairobi and the CSD Secretariat is based in New York. The Global Environment Facility is headquartered in Washington DC. The secretariat for the various global conventions are scattered: Montreal for the Convention on Biological Diversity; Bonn for the Climate Change Convention and Desertification Convention; Geneva for the Basel Convention and CITES. As conventions become more cross-referenced and entwined, something more than the existing strategies of 'bilateral' coordination between them becomes essential. UNEP itself has suggested that there is a 'compelling rationale for ... rationalizing, streamlining and consolidating the present system [of] multilateral environmental agreements' (cited in Biermann 2002, 300).

Konrad von Moltke has argued that the United Nations is 'famously resistant to any coordination' (cited in Thomas 2000, 13). At UNCED, calls for better coordination of environment and sustainable development policies and operations were generally welcomed although whether they were realistic is another matter. Ten years later, WSSD argued for 'limiting overlap and duplication of activities' (UN 2002, 65) and suggested that ECOSOC should 'increase its role in overseeing system-wide coordination' and that it should 'promote greater coordination, complementarity, effectiveness and efficiency of the activities of its functional commissions and other subsidiary bodies ... relevant to the implementation of Agenda 21' (UN 2002, 67).

Despite, or perhaps because of this enthusiasm, administrative coordination within the UN remains complicated and confusing and it is difficult to find any real evidence that it has improved in any meaningful or substantive way. Within a few months of UNCED a flurry of reform gave rise to the Department for Policy Coordination and Sustainable Development (DPCSD), the High-Level Advisory Board on Sustainable Development (HLAB) and the Inter-Agency Committee on Sustainable

Development (IACSD). DPCSD included the Division for Sustainable Development (DSD) which in turn was to act as the Secretariat for the CSD.[1] The IACSD was mandated to 'identify major policy issues and ensure effective system-wide cooperation and coordination' (see Flanders 1997, 392). The implementation of its functions, however, was turned over to a series of Task Managers involving almost *every* other organization or programme within the UN.

Following more reform in mid-1997, the Division for Sustainable Development (DSD) became part of the new Department of Economic and Social Affairs (DESA) which was itself formed from the merger of DPCSD with two other secretariat departments to achieve 'further streamlining and efficiencies'. Expectations within the UN were that the DSD would 'enhance the capacity of the Secretariat to support action at all levels to implement Agenda 21 and to ensure greater coherence in the Secretariat's work in the area of sustainable development in general' (UNGA 1997, para. 19), which rather suggests that earlier reforms were unsuccessful. Both DSD and UNEP participate in the Executive Committee on Economic and Social Affairs (ECESA), one of four new executive committees established to 'sharpen the organization by reducing duplication of effort and facilitating greater complementarity and coherence' (UN DESA 1999, n.d.).

In 1999, as noted above, the General Assembly approved the establishment of an Environmental Management Group, chaired by the Executive Director of UNEP, to improve inter-agency and inter-MEA coordination. In October 2000, the United Nations System Chief Executives Board for Coordination (CEB—the old Administrative Committee on Coordination) established a High Level Committee on Programmes to ensure coordination in programme areas including sustainable development. A year later, new administrative arrangements saw the abolition of the Inter-Agency Committee on Sustainable Development replaced by an Inter-Agency Meeting on Sustainable

---

[1]   Attempts to coordinate environmental concerns within the UN were not new. After the 1972 Stockholm Conference, the General Assembly established an Environment Coordination Board for this very purpose. The ECB was later replaced by the Administrative Committee on Coordination.

Development as part of a process to pursue coordination through more informal and flexible mechanisms. Suggestions that MEAs should be clustered (geographically or by environmental category) have made some headway in debates as a way of seeing coordination as a policy issue rather than a management one (see Oberthür 2002). Nevertheless, Mark Imber's observation that the UN's 'way of streamlining itself can move grown men [sic] to tears' (1994, 116) has not lost its relevance.

4   Multilateral Cooperation and Participatory Diplomacy

At one level, recent UN environmental diplomacy represents an extraordinary and perhaps unexpected degree of cooperation and institutional evolution and, as noted above, both UNEP and the General Assembly can take credit for this. Under UNGA resolutions or UNEP Governing Council directives, states have agreed to work together in a number of cooperative fora. Contemporary UN environmental diplomacy has its roots in the 1972 Stockholm Conference on the Human Environment. The number of multilateral environmental agreements adopted since 1972 has increased each decade, and most major environmental issues are now covered by some form of legally-binding treaty or at least a soft-law declaration which reinforces customary practice. The concept of sustainable development, which seeks to balance the imperatives of 'sustained economic development, improved social equity and environmental sustainability' (UNSG 1997a, para. 5) has been widely although not universally accepted as the primary objective of UN-sponsored environmental cooperation, negotiation and policy implementation. But the cooperative spirit has been limited: the environmental agreements adopted have reflected a lowest common denominator consensus; political and economic interests have taken precedence over environmental ones; tensions over issues of funding and technology remain unresolved in any of these agreements, despite recognition in almost all of them that this is a fundamental issue. At best, diplomatic processes are 'too slow [and] too laborious' (Razali cited in UN DPI 1996). At worst, they represent a 'total disconnect' with reality (Bramble cited in UN DPI 1996).

Nowhere was this contradiction more evident than at Rio. UNCED was charged, among other things, with strengthening international environmental cooperation. Maurice Strong, the Conference's Secretary-General, hoped that it would 'establish the basis for the new dimensions of international cooperation that will be required to ensure 'our common future' (Strong 1991, 297). In other words, UNCED was to herald a new era in international environmental governance and provide a model for post Cold War environmental cooperation and diplomacy. The conference was notable for the extensive participation of governments and political leaders who seemed prepared to contribute to the 'new and equitable global partnership' that the Rio Declaration claimed as its objective. But negotiations were dominated by a relative gains mentality, by last minute compromises reached in the final hours of the final days and by a general unwillingness of the industrialized countries to make a serious commitment to financial assistance to developing countries. The three agreements produced through two years of preparatory meetings and the two weeks of summit negotiations are all non-binding on the states who negotiated them: a statement of principles (the Rio Declaration), a programme of action (Agenda 21) and a statement of forest principles in lieu of any agreement about any kind of formal convention on managing forests or mitigating deforestation.

UNCED did, however, take an important step in broadening the parameters of participatory diplomacy, at least procedurally. Maurice Strong overcame objections from some governments to ensure that representation was not confined only to those NGOs with ECOSOC consultative status but would accommodate those who could demonstrate that their interests were directly relevant to the UNCED agenda.[2] As a result, 1400 NGOs were accredited to UNCED, with permission to participate in a range of working group meetings as well as plenary meetings. Other NGOs worked closely with friendly government delegations, including those from some smaller states whose expertise was

---

[2] The non-state actors present at UNCED and its preparatory process also included scientific organizations and groups representing various concerns within business and industry, local government and other 'stakeholders'.

stretched thinly compared with the large delegations from many of the richer (and more powerful) countries. Despite these efforts, participatory opportunities *decreased* as the preparatory process drew closer to completion. By PrepCom IV (in New York) UN security officers were deployed to ensure that NGO representatives were excluded from informal consultations, the small meetings at which many of the final decisions and compromises were reached (see Clark et al. 1998, 17; IISD 1997, 13). This exclusion from the 'informal-informals' continued at the summit in Rio itself.

UNCED mobilized a bout of frenetic negotiating activity on global environmental change which seemed to suggest that member states were energized by the Rio process, despite its limitations, and were prepared to move quickly to establish and perhaps even to implement legally-binding rules to mitigate environmental degradation. The two UN treaties tabled for signature at Rio were given international legal effect within two years of the conference. The Convention on Biological Diversity (CBD) came into effect at the end of December 1993 and the United Nations Framework Convention on Climate Change (FCCC) came into force in March 1994. The post-Rio agenda included negotiations for a UN Convention to Combat Desertification (CCD), opened for signature in October 1994 and ratified into international law in December 1996. It included negotiations for a UN Agreement on Straddling and Highly Migratory Fish Stocks (opened for signature on 4 December 1995) and the convening, in May 1994, of a global conference on the Sustainable Development of Small Island States.

While it might have been possible to justify some degree of optimism in June 1992, the UN's next major test of multilateral cooperation and participatory diplomacy—the General Assembly's 'Rio+5' Special Session in June 1997—demonstrated the fragility of the Rio spirit and suggested that while the *processes* might have been adjusting to the new imperatives of cooperation and inclusiveness, the content and outcomes had not. The imperatives for rejecting 'business as usual' diplomatic and management practices were made clear in two UN reports prepared for UNGASS.

The Secretary-General's report *Global Change and Sustainable Development: Critical Trends* drew attention to continued dangers associated with patterns of unsustainable development while holding out the possibility of positive and effective policy interventions (see UNSG

1997b). UNEP's *Global Environment Outlook* described several advances in institutional development, the implementation of sustainable development principles, attention to pollution and resource depletion and the increase in nongovernmental participation. Its blunt conclusion, however, was that 'from a global perspective the environment has continued to degrade ... [and] progress towards a sustainable future is just too slow' (UNEP 1997). UNEP's third GEO reinforced these concerns with its detailed retrospective, which concluded that 'overall, policy measures have not been adequate to counteract the pressures imposed by increasing poverty and uncontrolled consumption' (2002, 298).

The UNGASS outcomes were disappointing from both a diplomatic and environmental point of view. The Programme for Further Implementation was watered down in final negotiations. Many of the compromise outcomes on specific points were little different from previous agreements or 'no more ambitious than the positions officials had already identified as fallbacks' (Osborn and Bigg 1998, 12). The draft political statement had to be abandoned after it proved impossible to reach consensus. It was replaced with a six-paragraph statement of commitment inserted as a preamble to the Programme.

General Assembly President Ambassador Razali of Malaysia suggested that the fate of the political statement demonstrated that the nations of the world could not agree on how to work together to halt environmental degradation (cited in Rogers 1997). The problems were extensive but not new: continued disagreement over the meaning of sustainable development, let alone how to implement it; a general unwillingness to accept any new commitments especially with respect to finance and technology; confusion over how to incorporate the global economic agenda, including the impacts of globalization and the pursuit of trade liberalization. UK environment minister Michael Meacher was moved to describe the session as a 'chaotic and disconnecting experience' (cited in Jordan and Voisey 1998, 94). There is little to commend efforts at UNGASS, except perhaps to note that it helped to entrench NGO participation as a feature of UN environmental governance. For the first time at a General Assembly Special Session, NGOs and representatives of other major groups identified in Agenda 21 were able to deliver speeches to the Plenary and to have access to ministerial level consultations.

Even though the Special Session itself had achieved little, the general momentum of international environmental negotiation did not seem to falter. In December 1997, parties to the Climate Change Convention adopted the Kyoto Protocol which established greenhouse reduction targets for industrialized countries although the targets are limited and inadequate.[3] In September 1998, governments adopted the Rotterdam Convention on Prior Informed Consent for the international trade in a range of pesticides and industrial chemicals. In January 2000, governments agreed a biosafety protocol to the Biodiversity Convention (the Cartagena Protocol) and in December of the same year, the Stockholm Convention on persistent organic pollutants. In its Malmö Declaration, the Global Ministerial Environmental Forum anticipated that the World Summit on Sustainable Development in Johannesburg in August 2002 would 'reinvigorate international cooperation based on ... a spirit of international partnership and solidarity' (GMEF 2000, para. 1).

While the primary purpose of the WSSD was to accelerate implementation of Agenda 21 rather than to renegotiate the substance of the UNCED agreements, the General Assembly resolution which established the Summit also suggests a more broadly normative purpose as well. The meeting in Johannesburg was intended to *reinvigorate* at the highest political levels the global commitment to sustainable development, international solidarity and the North/South partnership (UNGA 2000, para. 17 (b)). The Commission on Sustainable Development again coordinated the preparatory process which involved an extensive round of multilateral consultations. Multi-stakeholder dialogues and regional roundtables of eminent persons were held in the middle of 2001. Five regional PrepCom meetings were held in the last four months of 2001 to examine issues of concern to each region and identify future priorities and the CSD convened four global preparatory committee meetings, chaired by Emil Salim, former environment minister of Indonesia.[4]

---

3   At time of writing, entry into force is still uncertain with US refusal to participate. The Protocol can still enter into force if Russia ratifies the agreement along with the many emitters which have already done so.

4   See Steiner (2003) for a brief discussion of the preparatory process.

Fears that WSSD would be little more than a 'conference to celebrate a conference' (cited in Brack, Calder and Dolun 2001, p. 1) were only partly alleviated. Two documents were adopted. The Political Declaration departs from the usual formality of statements of principle associated with the Stockholm and Rio conferences. It is presented as a 'solemn pledge' of collective responsibility to the peoples of the earth, to the 'children who represent our collective future' (UN 2002, 1-5). The WSSD Plan of Implementation, the second key document adopted in Johannesburg, includes a number of fairly broad commitments. Some repeat the Millennium Development Goals.[5] Others are watered down versions of what governments had adopted in other multilateral fora (see Pallemaerts 2003). There are a number of more specific commitments on what has become known as the WEHAB agenda—water, energy, health, agriculture and biodiversity.

The Johannesburg Plan of Implementation made a number of general statements about the institutional framework and the importance of good governance and a vibrant and effective UN system. It called for stronger collaboration within the UN system (UN 2002, para. 140(b)) and encouraged the General Assembly to consider the 'important but complex' issue of universal membership of the UNEP Governing Council (UN 2002, para. 140(d)). Under the Plan, governments also targeted CSD as the 'forum for consideration of issues related to the integration' of various dimensions of sustainable development and argued that the Commission should be 'strengthened' (UN 2002, para. 145) although it gave little guidance as to how this might happen.

WSSD did attempt to maintain an inclusive approach to debate and discussion, although non-government organizations and civil society groups were generally disappointed in the outcomes. It heralded the 'partnership initiatives', encouraging the negotiation of formal agreements on sustainable development programmes involving governments, international institutions, and a range of non-government and private sector actors. The Plan of Implementation, however, gave only a qualified

---

5   On the Millennium Development Goals, see UNDP 2003.

endorsement to principle 10 of the Rio Declaration which seeks to guarantee individuals access to environmental information, to decision-making and to judicial and administrative proceedings.

In practice, democratic governance *is* complex. UN diplomatic processes *have* become more inclusive and transparent, in accordance with the dictates of democratic pluralism as explained earlier in this chapter. Agenda 21 acknowledged nine major groups (sometimes referred to as the independent sector) as 'partners with Governments in the global implementation of the Rio agreements' (Sands 1995, xvii), bringing a range of stakeholder interests under the sustainable development umbrella. As Clark et al. observe, 'governmental frames have clearly been realigned in the 1990s to recognize a broader role for NGOs and some of their substantive innovations at global conferences' (1998, 33). NGOs have been influential in helping to define the environmental agenda, in pushing states to the negotiating table, in encouraging (and in some cases enforcing) implementation, in monitoring compliance and in demanding transparency. However, governments remain divided over the participation of environment and development NGOs, although they are often less exercised by the involvement of the business and industry sector.

The stakeholder approach to democratic governance within the UN system—that is, the more extensive inclusion of non-state actors—can overlook tensions among stakeholders and the ways in which participation is more easily available to some than to others. Participatory rights at multilateral negotiations, for example, have also become 'meaningful primarily for well-organized, well-financed and well-informed NGOs' (Anonymous 1991, 1589) and often more for northern-based NGOs than for those in the developing countries. Further, stakeholders have competing agendas that are not always congruent with the demands of sustainable development and environmental protection. The corporate sector has become an influential participant in UN environmental governance in a way that has been characterized as 'a marketing opportunity for large-scale business rather than a brake on the pursuit of profit maximization for its own sake' (Hughes and Wilkinson 2001, 157).

The social justice dictates of democratic governance, on the other hand, are rarely met. This approach requires that democratic governance (under the UN as elsewhere) recognizes the interests and voices of the most marginalized. It requires also that the outcomes of multilateral negotiations will ameliorate the inequities associated with global environmental change by which the poorest and most disadvantaged are often the most vulnerable to environmental change even though they have made almost no contribution to its causes. This requires more than bringing more voices to the negotiating table. Rather it requires a substantive commitment to providing assistance for poorer countries and people in their attempts to manage environmental degradation. Agreements such as the Climate Change Convention, the Desertification Convention and the Biodiversity Convention, the Rio Declaration and Agenda 21 recognize that women and indigenous people(s) should be included in decision-making. Some steps have been taken by some countries to give effect to these requirements and indigenous communities, for example, have also sought to invoke the provisions of agreements such as the Biodiversity Convention in claiming rights for traditional knowledge (see, for example, Fourmile 1998).

However, greater equity and environmental justice for poorer states and marginalized peoples has not been a consequence of any of the environmental agreements adopted under UN auspices. The general reluctance of industrialized countries to make any real commitment to providing sustainable development funding under UN or other auspices, their preference for foreign direct investment (which rarely goes to the poorest), the continued decline in OECD development assistance and long-standing and unresolved problems with non-payment of contributions to the various funding mechanisms established under multilateral environmental agreements all give evidence to the continued inequitable character of environmental governance. As the UNDP human development reports testify, not only is the gap between rich and poor being maintained but so too is the extent to which the latter remain the primary victims of environmental decline.

5   International Norms and Principles

The United Nations is an important source of international norms and rules. To the extent that it *has* occurred, the universalization of environmental norms, expressed in international law and state practice, has been closely tied to UN negotiations and agreements. Rather than challenging sovereignty, the general direction of normative reform within the United Nations has been to rethink the general rights and obligations of states. This is not surprising, given that membership of the organization is restricted, in effect, to nation-states.

Member states have proved generally unwilling to abrogate sovereign authority in the interests of a global environmental partnership and there is little evidence that the cosmopolitan norm of extraterritorial responsibility has become firmly embedded in state and inter-state practice. The most invoked commitment to the sovereignty norm is found in Principle 21 of the 1972 Stockholm Declaration that, in slightly modified form, became Principle 2 of the Rio Declaration. It asserts states' sovereign rights over their resources but also establishes responsibility for environmental damage beyond their borders (in effect the environment of other states) or areas beyond national jurisdiction such as the oceans. Statements of sovereign rights remain fundamental to most United Nations multilateral environmental agreements. Clauses affirming the *physical* rights of states to their resources and *authority* rights over how those resources are to be used and exploited appear, for example, in the preambles of the Vienna Convention for the Protection of the Ozone Layer, the Desertification Convention and the Climate Change Convention and in the body of the Statement of Forest Principles and the Biodiversity Convention.[6] This normative attachment to sovereignty has been reinforced rather than alleviated by continuing tensions between the developed and developing countries. In the absence of any firm commitments from the developed countries to help redress inequities in environmental cause and impact, and in the face of what they see as attempts to 'globalize' sovereign resources, developing countries have reasserted their authority over environment and development policy.

Efforts to manage this often difficult relationship between developed and developing countries have been embodied in the principle of common but differentiated responsibilities (CBDR). This principle attempts to meet the concerns of developed countries (particularly the United States)

---

[6]   Little progress has been made either on collective responsibility issues related to a states duty to inform and consult other states with respect to detrimental environmental impact of their activities or on the issue of legal remedies.

that all countries have obligations to address the global challenges of environmental degradation and unsustainable development and, at the same time, recognizes developing country concerns that those obligations are not the same for all countries. In particular, and the Rio Declaration (principle 7) and the climate change convention in article 3.1 articulate in explicit terms that industrialized countries have a responsibility to take the lead. In practice, however, CBDR has foundered on political differences over funding, targets and commitments.

Despite this determined attachment to sovereign rights in international environmental politics, new principles *have* fleshed out a potentially more robust normative framework for environmental governance under the United Nations. These new principles and norms serve to challenge traditional state-centric jurisdictions and widen the scope of those to whom obligations are owed in international law beyond states and beyond present generations. UN agreements now articulate the polluter pays principle (PPP), an anti-subsidy principle that demands the internalization of environmental costs although in more popular usage it has become a synonym for state and corporate liability. They have also served to internationalize the precautionary principle that requires that scientific uncertainty should not be grounds for postponing what the Rio Declaration calls 'cost effective measures', even if short-term political and economic interests dictate otherwise. The principle, which has been particularly relevant to the climate change debates, serves as a warning to states and non-state (usually corporate) actors in the face of uncertain futures. A third important principle and practice—that of prior informed consent (PIC)—has come to have particular resonance for developing countries in their attempts to resist the export of hazardous and toxic substances of various kinds from industrialized producers and governments. The first two of these three principles arose outside the UN process—PPP under the OECD and the precautionary principle initially from West German domestic law—but recognition within UN negotiations, and their articulation in the Rio Declaration as well as in agreements such as the convention on climate change or the biosafety protocol has given them international exposure although hardly universal force in practice.

New concepts such as the common heritage of humankind and intergenerational equity have also entered the international environmental lexicon although shaped as much by political as by normative or environmental imperatives. The common heritage principle, grounded in the UN Law of the Sea negotiations, suggests that common spaces, namely the high seas and outer space, should be managed for the good of all rather than on a first come, first served basis. While it would seem to articulate a norm for collective sovereignty and shared environmental stewardship (despite its genesis in political debates over common management of resource exploitation) there is little real evidence that it has been accepted in practice much beyond its application to the high seas.

Despite attempts to claim the atmosphere and climate system as a common heritage issue, the climate convention refers only to change in the Earth's climate as a common *concern* of humankind. Efforts to define biodiversity or forests as common heritage have met with little success, reflecting tension between ecological public goods and sovereign resources. The principle of intergenerational equity—that future generations have a right to inherit a planetary environment in at least as good a condition as previous generations have enjoyed—appears in the Stockholm Declaration, is fundamental to the articulation of sustainable development in the Brundtland Report and features in a number of multilateral environmental agreements. It is, therefore, a key UN principle although it is also contested and at times vague in its content (see Thompson 2001). Perhaps for these reasons the rhetoric is not matched by much in the way of operational practice. Certainly the generation which was immediately 'future' at the time of the 1972 Stockholm Conference is now rightly able to claim that its heritage has been undermined and reports such as UNEP's series of Global Environment Outlooks would suggest that the next generation will have similar grounds for complaint.

6   Conclusion

The United Nations is not, of course, a singular entity or unitary actor. In its most formal sense, it is an intergovernmental association of sovereign states and it is member states who must accept responsibility for UN outcomes. As Charnovitz points out, 'if governments wanted to make UNEP stronger now, they could do so' (2002, 15). Imber reminds us, however, that this does not mean that the UN is simply an instrument of state interests. It is also an institutional actor, through its various agencies and programmes, and an arena for the negotiation of international policy (Imber 1997, 218-22).[7] The assessment here confirms that the UN as both actor and negotiating arena has a mixed record in responding to the challenges of global environmental governance. It helps to explain, first, why proposals for a world environment organization are often entwined with concerns about the United Nations and, second, why there continues to be ambivalence about the UN as the site of such an organization. As Secretary-General Kofi Annan has observed, there remains a 'sizeable gap between aspiration and accomplishment' (UNSG 1997c, para. 4). One might suggest that the UN tries hard, but it could do better.

Despite many well-intentioned attempts at institutional reform, there is still no authoritative and well-resourced UN agency with overall policy and operational responsibility for environment and sustainable development of the kind anticipated by the Commission on Global Governance in the introduction to this chapter. Despite an extensive agenda of multilateral negotiations, cooperation and diplomatic outcomes are still measured as much in procedural terms as in substantive ones. Member states *have* cooperated under UN auspices to adopt a growing number of multilateral environmental agreements but they have not cooperated *sufficiently* to ensure that those agreements are effective, either in their provisions or their implementation.

Although the analysis here focused on processes rather than outcomes, it is clear that the problem—environmental degradation and unsustainable development—has not been eliminated or substantially ameliorated. Environmental diplomacy remains characterized by an unwillingness on the part of many member states to moderate particularistic national interest concerns for the common good, by mutual suspicion between developed and developing countries and by ambiguity over the role of NGOs and civil society organizations. The UN *has* mentored new principles that articulate, in theory at least, a more expansive and cosmopolitan normative framework. It has, however, proved consistently

---

7   UN Secretary-General Kofi Annan sought also to make this clear to member states prior to the Millennium Assembly (the 55th meeting of the General Assembly; see UNSG 2000, 5-6).

difficult to put those principles into practice and the normative framework remains constrained by what French President Jacques Chirac has called the 'determination of States to cling to an outmoded conception of sovereignty' (1999, 25).

There are limitations to the UN's multilateral capabilities and, contrary to the optimism of the Commission on Global Governance and the Secretary-General's Agenda for Reform, important questions must be asked about whether those limitations can be overcome. The UN's status and autonomy as an actor has always been constrained by the political interests of member states, particularly the most powerful among them, by budgetary restrictions and by what governments are prepared to allow it to do, or what they are prepared to commit to under its auspices. This does not demolish the UN's legitimacy, although it clearly complicates the issue enormously. Rather it suggests that more realistic expectations of the UN, and therefore of any world environment organization, might be required. Despite greater participation from civil society, and greater 'operational freedom' in a range of policy areas, the UN remains inter-governmental, not global. Internally, the United Nations remains the product of its history, subject to turf battles between its agencies and programmes where politics governs appointments as much as does merit, despite recent administrative and bureaucratic reforms. Rather than chastising the United Nations for having done so little, we should, perhaps, be surprised that it has achieved as much as it has.

References

Anonymous (1991), 'International Environmental Law', *Harvard Law Review*, vol. 104 (7), 1484–639.

Anonymous (1993), 'UNEP: 17th Governing Council', *Environmental Policy and Law*, vol. 23 (3–4), 118–42.

Anonymous (1995), 'Commission on Sustainable Development: Third Session', *Environmental Policy and Law,* vol. 25 (4–5), 163–77.

Biermann, F. (2002), 'Strengthening Green Global Governance in a Disparate World Society: Would a World Environment Organization Benefit the South?', *International Environmental Agreements: Politics, Law and Economics*, vol. 2, 297–315.

Blumenfeld, J. (1994), 'Institutions: The United Nations Commission on Sustainable Development', *Environment*, vol. 36 (10), 2–5; 33.

Brack, D., F. Calder and M. Dolun (2001), *From Rio to Johannesburg: The Earth Summit and Rio+10*, Briefing Paper no. 19, Royal Institute of International Affairs, Energy and Environment Programme, London.

Brenton, T. (1994), *The Greening of Machiavelli: The Evolution of International Environmental Politics*, Royal Institute of International Affairs, London.

Charnovitz, S. (2002), *A World Environment Organization*, United Nations University Institute of Advanced Studies, Tokyo.

Chirac, J. (1999), Speech of Jacques Chirac, President of the Republic of France, at the 50th Anniversary Meeting of the IUCN, [reproduced in] *Environmental Science and Technology*, vol. 33 (1), 24–7.

Chirac, J. (2002), Speech of Jacques Chirac, President of the Republic of France, at the World Summit on Sustainable Development, [reproduced in] *French Science and Technology*, no. 45 [2003], 3–4.

Clark, A. M., E.J. Friedman and K. Hochstetler (1998), 'The Sovereign Limits of Global Civil Society: A Comparison of NGO Participation in UN World Conferences on the Environment, Human Rights and Women', *World Politics*, vol. 51 (1), 1–35.

Commission on Global Governance (1995), *Our Global Neighbourhood*, Oxford University Press, Oxford.

Doran, P. (1996), 'The UN Commission on Sustainable Development, 1995', *Environmental Politics*, vol. 11 (2), Spring, 100–07.

Downie, D.L. and M.A. Levy (2000), 'The UN Environment Programme at a Turning Point: Options for Change', in P.S. Chasek (ed.), *The Global Environment in the Twenty-first Century: Prospects for International Cooperation*, The United Nations University Press, Tokyo.

Elliott, L. (2002), *Global Environmental (In)Equity and the Cosmopolitan Project*, CSGR Working Paper 95/02, University of Warwick Centre for the Study of Globalisation and Regionalisation, Coventry.

Elliott, L. (2004), *The Global Politics of the Environment*, 2nd edition, Palgrave Macmillan, London.

Flanders, L. (1997), 'The United Nations Department for Policy Coordination and Sustainable Development', *Global Environmental Change. Human and Policy Dimensions*, vol. 4 (7), 391–404.

Fourmile, H. (1998), 'Using Prior Informed Consent Procedures under the Convention on Biological Diversity to Protect Indigenous Traditional Ecological Knowledge and Natural Resource Rights', *Indigenous Law Bulletin*, vol. 4 (14), 14–17.

French, H.F. (1995), *Partnership for the Planet: An Environmental Agenda for the United Nations*, Worldwatch Paper no. 107, Worldwatch Institute, Washington DC.

Gleeson, B. and N. Low, (2001) (eds), *Governing for the Environment: Global Problems, Ethics and Democracy*, Palgrave Macmillan, London.

Global Ministerial Environmental Forum (GMEF) (2000), *Malmö Ministerial Declaration*, UNEP, Nairobi.

Held, D. (1996), *Models of Democracy*, Stanford University Press, Stanford.

Hughes, S. and R. Wilkinson (2001), 'The Global Compact: Promoting Corporate Responsibility?', *Environmental Politics*, vol. 10 (1), 155–59.

Imber, M. (1994), *Environment, Security and UN Reform*, Macmillan, London.

Imber, M. (1997), 'Geogovernance without Democracy? Reforming the UN System', in A. McGrew (ed.), *The Transformation of Democracy?*, The Open University Press, Milton Keynes.

Imber, M. (1999), 'The Impasse in UN Reform', *Global Environmental Change. Human and Policy Dimensions*, vol. 9 (4), 329–32.

IISD—International Institute for Sustainable Development (1997), 'Summary of the Nineteenth United Nations Special Session to Review Implementation of Agenda 21', *Earth Negotiations Bulletin* vol. 5 (88), 1.

IISD—International Institute for Sustainable Development (1999), 'CSD–7: Briefing for Tuesday 20 April' [http://www.iisd.ca/linkages/csd/csd7/; accessed 21 April 1999].

Joint Declaration (1997), *Joint Declaration for a Global Initiative on Sustainable Development*, issued by the Department of Foreign Affairs, Pretoria, 23 June [available at http://www.polity.org.za/html/govdocs/pr/1997/pr0623c.html?rebookmark=1]

Jordan, A. and H. Voisey (1998), 'The Rio Process: The Politics and Substantive Outcomes of "Earth Summit II"', *Global Environmental Change. Human and Policy Dimensions*, vol. 8 (1), 93–97.

Khor, M. (1994), 'The Commission on Sustainable Development: Paper Tiger or Agency to Save the Earth?', in H.O. Bergesen and G. Parmann (eds), *Green Globe Yearbook 1994*, Oxford University Press, Oxford.

Kirton, J. (2005), 'Generating Effective Global Environmental Governance: The North's Need for a WEO', in F. Biermann and S. Bauer (eds), *A World Environment Organization. Solution or Threat for Effective International Environmental Governance?*, Ashgate, Aldershot, 145–172.

Knight, W. A. (1995), 'Beyond the UN System? Critical Perspectives on Global Governance and Multilateral Evolution', *Global Governance*, vol. 1 (2), 229–53.

Litfin, K.T., ed. (1998), *The Greening of Sovereignty in World Politics*, The MIT Press, Cambridge (Mass.).

Lodefalk, M. and J. Whalley (2002), 'Reviewing Proposals for a World Environmental Organisation', *The World Economy*, vol. 25 (5), 601–17.

Newell, P. (2002), 'A World Environment Organisation: The Wrong Solution to the Wrong Problem', *The World Economy*, vol. 25 (5), May, 659–71.

Oberthür, S. (2002), *Clustering of Multilateral Environmental Agreements: Potentials and Limitations*, Working Paper, United Nations University Institute of Advanced Studies, Tokyo

Osborn, D. and T. Bigg (1998), *Earth Summit II: Outcomes and Analysis*, Earthscan, London.

Pallemaerts, M. (2003), 'International Law and Sustainable Development: Any Progress in Johannesburg', *Review of European Community and International Environmental Law*, vol. 12 (1), 1–11.

Rogers, A. (1997), 'Earth Summit+5 Talks Run Into Impasse at 11th hour', WETV/Webcast, [available at http://www.SustainableDevelopment.net/empire/?SubSystemID=2&-ComponentID=332]

Sandbrook, R. (1999), 'New Hopes for the United Nations Environment Programme (UNEP)?', *Global Environmental Change. Human and Policy Dimensions*, vol. 9 (2), 171–74.

Sands, P. (1995), *Principles of International Law I: Frameworks, Standards and Implementation*, Manchester University Press, Manchester.

Steiner, M. (2003) 'NGO reflections on the World Summit: Rio+10 or Rio–10?', *Review of European Community and International Environmental Law*, vol. 12 (1), 33–38.

Strong, M. (1973), 'Introduction', in W. Rowland, *The Plot to Save the World: The Life and Times of the Stockholm Conference on the Human Environment*, Clarke, Irwin and Co., Toronto.

Strong, M. (1991), 'ECO '92: Critical Challenges and Global Solutions', *Journal of International Affairs*, vol. 44 (2), 287–300.

Thomas, U. (2000), 'Improving Integration between the WTO and the UN System', *Bridges*, vol. 4 (8), 13–14.

Thompson, J. (2001), 'Planetary Citizenship: Definition and Defence of an Ideal' in B. Gleeson and N. Low (eds), *Governing for the Environment: Global Problems, Ethics and Democracy*, Palgrave, Basingstoke.

Töpfer, K. (1998) 'United Nations Task Force on Environment and Human Settlements', *Linkages Journal*, vol. 3 (3); [available at www.iisd.ca/linkages/journal/toepfer.html].

UN [United Nations] (2002), *Report of the World Summit on Sustainable Development*, A/CONF.199/20, United Nations, New York.

UNCED [United Nations Conference on Environment and Development] (1992a), *Report of the UN Conference on Environment and Development: Annex I, Rio Declaration on Environment and Development*, A/CONF.151/26 (vol. 1), 12 August.

UNCED (1992b), *Report of the UN Conference on Environment and Development: Annex II, Agenda 21*, A/CONF.151/26 (vol. I–III), 12 August.

UN DESA [United Nations Department of Economic and Social Affairs] (1999), *Executive Committee on Economic and Social Affairs*; [available at http://www.un.org/esa/-coordination/ecesa/ecesa.htm;accessed 9 August 1999].

UN DPI [United Nations Department of Public Information] (1996), 'Earth Summit+5', DPI/SD/1864, December 1996; [http://www.un.org/ecosocdev/geninfo/sustdev/-es&5broc.htm; accessed 5 March 1998].

UNDP [United Nations Development Programme] (2003), *Human Development Report 2003—Millennium Development Goals: a Compact Among Nations to End Human Poverty*, Oxford University Press, New York.

UNEP [United Nations Environment Programme] (1997), *Global Environment Outlook: Executive Summary*, Nairobi: UNEP; [http://www.uneorg/unep/eia/geo1/exsum/ex2.-htm; accessed 10 March 1998].

UNEP (1999a), *UNEP: Guardian of the Global Environment*, News Release 1999/13; [http:/www.uneorg/unep/per/ipa/pressrel/r02-0699.001; accessed 16 August 1999].

UNEP (1999b), *Global Environment Outlook 2000*, Earthscan, London.

UNEP (2002) *Global Environment Outlook 3*, Earthscan, London.

UNGA [United Nations General Assembly] (1989), *United Nations Conference on Environment and Development*, Resolution 44/228, 85th Plenary Meeting, 22 December.

UNGA (1997), *Environment and Sustainable Development: Special Session for the Purpose of an Overall Review and Appraisal of the Implementation of Agenda 21—Outcome of the Nineteenth Special Session of the General Assembly: Report of the Secretary-General*, A/52/280, 14 August.

UNGA (1999), *Report of the Secretary-General on Environment and Human Settlements*, A/RES/53/242, 10 August.

UNGA (2000), *Ten-Year Review of Progress Achieved in the Implementation of the Outcome of the United Nations Conference on Environment and Development*, A/RES/55/199, 20 December.

UNSG [United Nations Secretary-General] (1997a), *Overall Progress Achieved Since the United Nations Conference on Environment and Development: Report of the Secretary-General*, E/CN.17/1997/2, 31 January.

UNSG (1997b), *Global Change and Sustainable Development: Critical Trends*, Report of the Secretary-General, Commission on Sustainable Development, Fifth Session [http://www.-un.org/dpcsd/dsd/trends.htm; accessed 5 March 1998].

UNSG (1997c), *Renewing the United Nations: An Agenda for Reform*, A/51/950, 14 July.

UNSG (2000), *We the Peoples: The Role of the United Nations in the 21st Century*, United Nations Department of Public Information, New York.

UN Task Force on Environment and Human Settlements (1999) *Report*, annexed to Report of the Secretary General on Environment and Human Settlements, General Assembly, A/53/463.

Vellinga, P., R. Howarth and J. Gupta (2002), 'Improving Global Environmental Governance', *International Environmental Agreements*, vol. 2 (4), 293–96.

Wapner, P. (2003), 'World Summit on Sustainable Development: Toward a Post-Jo'burg Environmentalism', *Global Environmental Politics*, vol. 3 (1), 1–10.

WCED [World Commission on Environment and Development] (1987), *Our Common Future*, Oxford University Press, Oxford.

Chapter 3

Global Environmental Governance: Challenges for the South from a Theoretical Perspective

Joyeeta Gupta*

1   Introduction

With the globalization of the economy and the transboundary nature of many environmental problems, the issue of how best to design global environmental governance systems has taken centre-stage—and the debate on a world environment organization represented in this volume is just one part of the larger discussion.[1] Of course, the word 'design' suggests that some theorist in his/her ivory tower can craft a system that will work most effectively. This would ignore the fact that global systems are the result of protracted and complex negotiations, and that design issues may influence the process in the sense that those negotiating may take these aspects into account, but at the same time, they will negotiate on the basis of their 'national' interests in relation to the specific problem.

Again, in general, most design specialists have tended to focus on how best environmental interests, as perceived by the design specialists,

---

*   This chapter is written as part of the project 'Inter-governmental and private environmental regimes and compatibility with good governance' financed by the Netherlands Scientific Organization.
[1]   See e.g. Agarwal et al. 1999; Keohane et al. 1993; von Moltke 2002; Gupta 2002a; Biermann 2002; Oberthür 2002; Young 2002.

can be dealt with, or how best the design meets criteria and principles that are of vital interest in the developed world. There is an implicit assumption that such design criteria would also somehow deal with the problems, priorities and characteristics of developing countries.

Against this background, this chapter specifically examines the question of how developing countries are affected by the diverse governance options and what the coping strategies are that they need to adopt to deal with these governance options in the context of globalization. In order to address this question, this chapter first presents the range of governance options, their critical features and their feasibility from different perspectives. It then examines the critical issues facing the developing countries. It brings the two sections together in an analysis of how different options will affect developing countries. It finally elaborates on how developing countries can improve their situation and the issues they need to take into account while participating in such negotiations.

A few notes of caution are not out of place here. In general, the group of developing countries refers to about 130 G-77 members. After the fall of the Berlin Wall, some of the East European countries were invited to participate as part of the developed world, while others like Kazakhstan became de facto part of the developing world. At the margin, there are developing countries that are very rich (e.g. Singapore); this group is not the main target of the analysis. There are also large rapidly developing countries like China and India—but both countries despite their rapid growth rate still have low per capita incomes and fall into the group of developing countries. Although there are large differences between the developing countries, there are geographical, historical, cultural, political, developmental similarities that are structural.[2] Furthermore, by virtue of the fact that the developed world, including the enlarged European Union is a strong bloc, the rest of the world has been acting as one group. Furthermore, this chapter takes an ideal-typical approach to clarify the

---

[2]   See for a detailed analysis of similarities and differences in the negotiating arena, Gupta 2003a.

challenges facing the generic developing country, but in doing so it may loose out on the nuances.[3]

2 Governance Options from a Theoretical Perspective

As a first step to the analysis offered in this chapter, it highlights the different governance arenas visible today. Then it presents a list of the different governance options and their key characteristics. Finally, it presents an analysis of the feasibility of these options in relation to different schools of thought.[4]

New Governance Arenas

In the process of globalization, two arenas can be distinguished—managed globalization and spontaneous globalization. Managed globalization refers to the attempts of states to control the processes of globalization as well as to institutions and processes that try to follow the principles of the rule of law and of good governance, and use the interstate treaty negotiation process based on jointly negotiated rules of procedure and the law of treaties. Spontaneous globalization refers to the processes by which the spread of knowledge and information (through the internet, world-wide web and global media), products, technological and scientific innovation (through global markets), financial and economic integration (through financial systems), crime, corruption and terrorism create new rules explicitly or implicitly for the functioning of the global economic and political order. Although even spontaneous globalization is embedded within the framework of international law in that it is not illegal per se and since international law covers all legal phenomena everywhere (Allot 1999),

---

3 Much of the theory in this chapter is based on an inductive analysis of more than 600 interviews conducted mostly with developing country stakeholders over the last ten years and literature coming from these countries on, but not limited to, the international negotiations on climate change.

4 This section draws on Gupta 2002a.

it in fact refers to institutions and processes that develop as a result of scientific and technological innovation and societal promotion which are often in advance of interstate rules and regulations that can set the boundary conditions of what is acceptable and what is not.

With globalization, a range of different governance relationships is becoming visible (IDGEC 1999, Kersbergen and Waarden 2001, Young 2002). In addition to horizontal governance relationships (that include treaty negotiations and political declarations at different administrative levels from local to global levels) and vertical governance (multi-layered governance involving the interactions between different administrative levels), there is increasing evidence of parallel governance relationships. This latter refers to self-regulation by non-state actors, codes of conduct, voluntary labelling and certification schemes (Bendell 2000; Gibson 1999; Mannheim 1999, Campins-Eritja and Gupta 2002). There are also diagonal governance relationships involving public-private partnerships and partnerships between non-state actors and state actors and international organizations such as the so-called 'type 2' outcomes being promoted by the World Summit on Sustainable Development (von Moltke 2002, Gupta 2003b).

Different Governance Options

An examination of the literature on global governance options reveals that in essence many focus on the areas of governance in the arena of managed globalization (Biermann and Simonis 1998, Biermann, this volume; Schrijver and Weiss 2002; Esty 1994; ILA 2002; Oberthür 2002), while there are also others who focus more on governance options in the arena of spontaneous globalization (von Moltke 2002; Snyder 2001).

Some would like to see a hierarchical integrated body such as a World Sustainable Development Organization, high up in the UN hierarchy, which has a mandate, the power and resources to steer decision-making throughout the UN system to promote sustainable development. Such a body would of necessity involve the mergers and take-overs of different existing UN bodies. Such a body would be higher than the World Trade Organization, occupying a position akin to the Security Council.

Some would like to see a hierarchical but single-issue body such as a World Environment Organization. Such a body would also be high up in the UN hierarchy and have a competing position to the World Trade

Organization and possibly the Bretton Woods Institutions. The major environmental organizations and regimes within the UN system would be amalgamated within such a body, and this would ensure internal consistency, efficiency and environmental effectiveness within the body and a strong, coordinated, continuing challenge to actions taken by other powerful UN bodies that affect the environment negatively.

Others call for a non-hierarchical focal point—a mere amalgamation of some UN agencies and regimes to ensure improvements within the system through better coordination (see for example Biermann and Charnovitz, this volume). Yet others, including the organizations themselves, call for a strengthening of their mandate and powers so that they can function effectively. Some argue that such reorganization would be vastly expensive and impractical and instead it would make sense to simply have a high-level advisory group that followed developments possibly both in the field of managed and spontaneous governance and could advise the Secretary General to make strategic interventions where it was seen necessary to promote sustainable development.

A more recent call is to promote the progressive development of the law of sustainable development. The idea behind this is that once the principles are identified, clustered and codified, they may begin to influence state behaviour through their normative force, and as they do so, state practice and that of international organizations will increasingly reflect these principles and these principles will gradually enter into the realm of customary international law and may even become obligations *erga omnes*.

One of the more pragmatic options is that of clustering the environmental regimes, and/ or some of the functions within these regimes in order to gain efficiency (see von Moltke, this volume). In the area of managed governance, some argue that the actual decisions are not taken in regimes and their formal processes, but in networks, bureaucratic or scientific, and argue that since these networks make the decisions, attempts to improve global governance should focus on improving these networks. Those researchers who believe that managed governance regimes are marginal in terms of their effect tend to focus more on the self-regulation initiatives within the world of non-state actors and argue that in the final analysis these explicit and implicit self-regulatory initiatives determine environmental outcomes and hence the focus should be on, if not, controlling these initiatives, understanding them and learning to cope with them.

The Feasibility of These Options from Different Schools of Thought

It is clear that not all these options are feasible. One can extrapolate from the perspectives of different schools of thought about the likelihood of the different options (cf. table 1 on p. 63). Four distinct schools of thought can be identified. The neorealist school of thought sees the state as a central actor in managed governance and believes that power politics strongly influences negotiation outcomes (e.g. Grieco 1996). From the perspective of this school of thought, a hierarchical integrated WSDO is impossible, and a WEO with strong or weak powers is highly unlikely, because states are unlikely to want to surrender their sovereignty on issues to a more or less powerful supranational or international body; and because international organizations are unlikely to support such a move. The development of an advisory body or clustering of regimes and their functions is more likely to be feasible. Focusing on improving network processes and on understanding and controlling the spontaneous governance processes may yield more results. Promoting the law of sustainable development may be feasible but is seen as irrelevant as states are not necessarily bound to the international treaties, and international rules 'may exist, but they do not exert an independent influence on state behaviour' (Arend 1996: 289).

The historical materialist school of thought tends to focus on power differentials and the asymmetry of the global negotiating blocs and the tendency for third world interests to get marginalized in the process of negotiation. Historical materialists tend to agree with neorealists, except that they believe that if the neorealists are wrong, and powerful supranational bodies can be set up, they believe that these bodies will not promote the interests of the periphery, and that these countries will become more and more marginalized in the process. The neoliberal institutionalist school of thought recognizes the importance of power, but believes that power theories alone cannot explain the developments in the international arena (e.g. Keohane 1996). Its scholars argue that power configurations differ from issue to issue and in benign issue areas, it may be possible to secure environmental agreement. While neoliberal institutionalists are sceptical about the possibility of integrating all the different issues into one body, they see the possibility for developing issue-linkages and clustering different options together; and they also see promise in focusing on decentralized networks.

Table 1. Global Environmental Governance Options

Design Options	Features	Feasibility			
		NR	HM	NI	IS
Hierarchical, integrated (e.g. WSDO)	Situated very high in the UN hierarchy with the power and mandate to steer all UN agencies	--	-	--	+
Hierarchical, single issue (e.g. WEO)	Situated high in the UN hierarchy with concentrated environmental responsibilities and the influence to steer UN agencies in relation to environmental issues	-	-	-	+
High Level Advisory Group	A highly effective, but very small and focused and hence relatively inexpensive group of advisors that signal environmental challenges and potential solutions	++	++	++	++
Non-hierarchical focal point	A world environment organization with a wide environmental mandate resulting from the mergers to takeovers of several environmental organizations, but without much influence or power over other bodies.	-	-	+	+
Strengthening individual organizations	This calls for no re-organization of the existing system, but for strengthening those institutions that perform well and are able to deal with the serious challenges ahead.	+	+	+	++
Promoting coordination through common principles (e.g. the Law of Sustainable Development)	This calls for no reorganization, but for the development of a code of conduct in relation to environmental and developmental issues throughout the UN system by promoting the progressive development of international law	0	+	0+	++
Regime clustering	This calls for no re-organization but the clustering of regimes or functions in order to make implementation more effective	--	+	+	++
Decentralized network organizations	This calls for more focus on the ways in which decentralized network organizations and processes influence policymaking	++	++	++	+
Multiple, pluralistic, horizontal, vertical, parallel and diagonal competing, and sometimes consistent regulatory frameworks		++	++	+	+

Code: NR: Neo-realism, HM: Historical materialism, NI: Neo-Institutionalism, IS: Idealistic Supranationalists; --: Impossible; - Unlikely; 0: Irrelevant; +: Necessary; ++ Feasible.
Source: Derived from Gupta 2002a, Tables 1-4.

The idealistic supranationalist school of thought reasons from an assessment of global problems, arguing that if global problems are to be addressed, they need to be dealt with within the framework of a supranational body with strong powers and mandate and which can stand above narrow national interests. Idealistic supranationalists would argue strongly in favour of a global constitution (e.g. Allott 2001), or the development of an international law of sustainable development at the very least (Schrijver and Weiss 2002). Another variant of this family would focus on promoting a strongly integrated hierarchical body based on democratic principles and the rule of law. They would possibly see the options of decentralized network organizations and self-regulatory frameworks as marginal issues, if the centralization process were successful.

Governance Options: Scenarios for the Future

Let us now turn to what is likely to happen in the evolution of governance patterns. This evolution is based on the initial assessment of feasibility of the different options listed in Table 2. I have identified four possible medium-term end goals in global environmental governance that correspond to, but are not necessarily entirely compatible with the theoretical approaches.

Table 2. Scenarios of Governance

Scenarios	End situation
Neorealist	Multiple competing small regimes; a free-market of governance options; weak regimes are rooted out to create efficiency in the system. Bureaucratic and scientific networking may become dominant.
Historical materialist	Most likely multiple competing regimes; but even if a strongly coordinated option emerges, Southern interests are likely to be marginalized
Neoinstitutionalist	Clustered regimes or functions to increase efficiency
Idealistic supra-nationalist, Variant 1	Strong vertically integrated centralized coordination of global environmental governance issues; based on good governance principles
Idealistic supra-nationalist, Variant 2	Progressive development of the international law of sustainable development that permeates all activities

In other words, the options listed above will possibly compete with each other, and may lead to five dominant outcomes. Although a strong vertically integrated centralized organization appears necessary from an analysis of the environmental problem, its likelihood is extremely low. Rather some regimes will be clustered, some will be strengthened and there will be competition between these regimes. It is also possible that this is accompanied by the progressive development of international law in the area of sustainable development.

3 Theoretical Approaches to Developing Country Participation

This section addresses the question of developing countries in the context of global environmental governance. It first defines them and provides a brief historical context of the way they participate in international decision-making. It then examines the key theories that explain the nature of developing country participation in environmental treaties. The 'developing countries' are a group of about 150 countries consisting of around 130 members of the Group of 77 and around 20 new countries mostly from the former East Bloc that by virtue of not belonging to the developed world are considered a part of the developing world by default. The bulk of these countries share a common history of colonialism and post-colonial attempts at rebuilding their own economies. They have weak economies, soft currencies, and often are politically unstable. Most of these countries are wrestling with the task of providing basic necessities to their people. Often, they exploit their environments in order to be able to produce commodities for the international market. They share many characteristics and since 1964, they negotiate together under the umbrella of the 'Group of 77' (cf. Nyerere et al. 1990; Agarwal et al. 1992; Ramphal 1983).

In the area of global environmental governance, developing countries are often invited to participate in international negotiations on a number of complex environmental issues. The issues that normally reach the international arena are those that dominate western domestic agendas. Thus, developing countries often are ill prepared to negotiate these issues. This does not imply that on occasion, individual countries and negotiators do negotiate extremely effectively. It merely implies that during the bulk of the negotiations, there are major challenges facing developing countries.

Drawing on the case of climate change negotiations I have identified eight sets of cumulative hypotheses that one can make about how developing countries negotiate on issues that are first signalled by Western countries and then imported to the domestic agenda of developing countries.5 These are:

(1) The Hollow Mandate. Most developing countries have a hollow mandate for negotiating global environmental issues (which implies a framework of ideas, but not much detailed content). This hollow mandate is a result of domestic ideological dilemmas, a structural imbalance in the knowledge available domestically and internationally, a focus on making political or bargaining issue-linkages rather than effective material issue-linkages and a tendency to define national interests in qualitative, elitist and diplomatic terms. They are also the result of domestic procedures that, if existent, are mostly meetings of form rather than substance. These procedures are affected by 'the four Ds': the delegation of responsibility to one ministry to deal with the issues, while the more important domestic ministries do not get constructively involved in the process; the diminution of responsibility with the delegated ministry being implicitly or explicitly given the powers to only discuss matters within its purview; the downsizing of the delegation to one or two negotiators; and discontinuity in the delegation, which implies that the delegates are regularly changed and there is not much institutional memory of the negotiating process.

(2) The Defensive Negotiating Strategy. The negotiation strategy tends to be defensive consisting of ad libbing, lacking substantive proposals, with a focus on opposition of the proposals of others, a tendency to vacillate when side-payments are made, a focus on principles and holistic approaches, and a tendency to be resentful of the negotiation outcomes at the end of the negotiations. In general one can argue that the strategy is to ensure that the national position expressed is somewhat legitimate, by basing it on proxy indicators of legitimacy. These proxy indicators include compatibility with

---

5   This section draws on Gupta (1997, 1999, 2000a, 2000b, 2000c, 2001a and 2002b).

precedents, principles and positions in other issue areas, even if these are inappropriate for the problem at hand.

(3) The Handicapped Coalition Forming Power. Although, as mentioned above, the G-77 is an organized negotiating body for the developing countries, the reason that they have handicapped coalition forming power is because pooling the individual hollow mandates of the developing countries does not necessarily lead to more information at G-77 level and hence a better strategy. Instead, given the physical disabilities such as poor communication facilities, the number of languages spoken or the poor political relations between many Southern countries, the situation is not yet ripe for them to pool their knowledge in such a way that the sum of their combined knowledge and position is higher than the sum of their individual positions. Instead, frequently, they are negotiating in an information vacuum, they lack staying power in negotiating specific issues, they do not trust each other enough to share responsibilities, their interests are best articulated in abstract, vague and rhetorical terms, and all this essentially adds up to handicapped coalition forming power.

(4) The Handicapped Negotiating Power. The strategy of individual countries is defensive. The negotiating strategy of the group is not just collectively defensive but also brittle, because countries can be bought off with side-payments, and threadbare, because there are many weak spots in the negotiating strategy.

(5) The Structural Imbalance in Bargaining. When developing countries negotiate with developed countries, they are not only confronted with their own negotiating weaknesses, they are also confronted with strong negotiating strategies and the political bargaining power of the developed countries. This implies that mostly when the developed countries are constructive and the developing countries defensive, this leads to coercion, concessions, partial problem solving and decision-less decisions. On the few occasions that developing countries are constructive, there is a strong likelihood that the developed countries will be defensive leading to exclusion, accommodation on paper and non-decisions. In many global issues, both are defensive leading to symbolic decisions, avoidance and indeterminate language. Only if both are constructive is there an

opportunity for problem-solving or possibly protracted conflict resolved through institutional learning (Gupta 2000a).

(6) The Competing Hypotheses of Problem-Solving. To some extent, theories on problem-solving can help to explain the North-South stress in global environmental issues. Realists have always argued that the powerful countries will not be willing to address problems if the solutions go against their national interests. Regime analysts have pointed out that despite the realist prediction there are examples of cooperation in difficult areas—and this is because issue-related power differs from problem to problem and in the complex web of decision-making there is space for cooperation. Such cooperation tends to be possible when the sub-issues are structured and benign, in other words when the science is uncontroversial and the values shared. However, science is often controversial and the values are not shared and when countries enter into negotiations pretending that the problem is structured, this gives rise to protracted controversy. In such situations it is best to invest in dialogue and social learning, rather than negotiating, as a strategy in order to be able to reach to constructive bargaining (Hisschemöller and Gupta 1999).

(7) Decreasing Legitimacy in North-South Negotiations. When the negotiating process proceeds with regularity and issues become more complex, developing countries get weaker in the negotiating process as time proceeds, because with each round they fall a few steps back. Thus, the negotiating outcomes rarely represent the positions of developing countries and are likely to become less legitimate over time (Gupta 2001a).

(8) Regulatory Competition and Late-Comers. The inability of developing countries to negotiate effectively on issues signalled by the North is aggravated by the attempt of developed countries to engage in regulatory competition in order to upload their domestic policies on to the international arena so that the costs of implementation are the least for them. This implies that the systems developed at the international level are likely to impose high administrative, legal and institutional costs on the developing countries when they have to implement these systems (see Börzel and Gupta 2000; Gupta 2004 forthcoming; Héritier, Knill and Mingers 1996).

The above has described sets of hypotheses that are likely to be valid when developing countries negotiate in relation to issues or instruments signalled by the North. However, situations may occur where the issues signalled by the South reach the global agenda, too. Take, for instance, the New International Economic Order, which reached the global agenda in the 1970s as a priority of the developing countries.[6] Also, in 1992, the developing countries were able to table many of their priorities in Agenda 21. But the financial mechanism that was eventually developed did not finance these issues, and instead focused initially on global issues and later on focal issues (Reed 1993; Gupta 1995). In 2002, the World Summit on Sustainable Development attempted at tabling issues of concern for both developing and developed countries. But the Type 1 agreement—the inter-state agreement—ended up eventually as a Declaration and Plan of Action, and even to this soft law instrument, the United States saw fit to add a reservation stating that the Declaration in no way implied any new obligations under international law (WSSD 2002). The Type 2 outcomes promoted at the Conference are typical examples of diagonal governance attempts; but there are potentially a number of problems that face such agreements (Gupta 2003b).

The above shortcomings apply primarily in the arena of managed governance. In the arena of spontaneous governance, there is little literature and documentation available regarding developing countries' ability to influence these processes. There is, however, considerable concern in the few documents available about their inability to influence the negotiating processes leading to, for example, the adoption of eco-labelling schemes in the West (e.g. Nath 1997; Jha 1997; Bharucha 1997; Campins-Eritja and Gupta 2002).

At one point, it was concluded that the development of a 'Group of 15' developing countries could help to structure developing country concerns and could lead to a more effective negotiating strategy for the South. However, Sridharan (1998, 370) concludes that: 'It would appear

---

6 Unfortunately, the negotiated instruments were never implemented by industrialized countries (Schrijver 1995).

that the G15 mirrors the dilemmas of the larger Third World bodies which are unable to evolve an unambiguous position on key issues because of their heterogeneity and their different levels of development, which impede joint action'. All this does not imply that on individual occasions developing country negotiators are not able to get a very good deal for themselves. But that is more the exception than the rule.

4   Governance Options: Challenges for Developing Countries

This section examines the policy question: How should developing countries prepare for a changing institutional framework at the global level? In order to do so, it will first examine which of the governance options are of interest to the developing countries. It then highlights some of the international negotiation trends and follows up with focused recommendations about how developing countries should prepare for international negotiations in global environmental governance.

Acceptability for Developing Countries

The big problem for developing countries is coping with the multiple initiatives and the tendency of the developed world to indulge in forum shopping. This leads some authors such as Biermann (2002) to conclude that a strong world environment organization would be most beneficial to developing countries, since they can pool their meagre resources together and present a common and effective front. While the argument in favour of a world environment organization is very persuasive in its content there are two limitations one needs to be aware of. The first is what I believe to be a structural problem in the South, namely its inability to access and pool its available resources (see Gupta 2000b). This can be seen not only in the way the G-77 and G-15 negotiate, but the state of the G-77 office and its records and publications and the limited influence of the South Centre (see South Centre 1993).

Second, although this to some extent implies that the fragmentation of efforts is minimized and the opportunities for forum shopping lost, the fact will remain that the North will become even stronger if it can focus the strength of its available resources in one forum. This may give the impression that there will be considerable efficiency gains, but I argue that such a structuring process will leave no room for alternative thought processes which might work against the interests of the South. If a majority voting mechanism is developed in this forum—or indeed a world environment organization—it might give developing countries more power; but this is extremely unlikely. Voting mechanisms are increasingly moving towards consensus and double majority rules. Given this, although there are clearly some advantages for developing countries, unless the global system eventually mirrors the ideal domestic system with a separation of powers between administration, legislation and judiciary, focusing exclusively on this option is a risky strategy because it may either back-fire against developing country interests, or because it is simply not a realistic option. This does not mean, however, that those supporting such an international organization are wrong; it merely means that developing countries cannot focus exclusively on this option.

In the case of a high-level advisory body, the developing countries may sometimes strike lucky in that those selected to this position may be more sympathetic to their perspectives. In any case, they will have to actively represent their case in such situations. History records that when individual organizations were sensitive to the views of the developing countries—as, for instance, UNCTAD, UNEP or WIPO—they have often received less financial resources and have become marginalized in the international arena (Gosovic 1992). There is no reason to suppose that this will not continue to happen into the future. In the area of the progressive development of international law there is more reason to suppose that the work of jurists may provide the doctrinal support and such support may be independent of political power, but may at the initial stages have less influence. Networks generally accelerate learning processes; but if there is little input from developing countries, learning will be a one-way process and will negatively affect developing countries.

Global Trends

Let me now sum up eight trends that are visible in the global arena. These trends are likely to dominate, irrespective of the position (or lack of it) of developing countries.

(1) Rapid scientific evolution. There is rapid scientific evolution in the developed countries facilitated by developments in technology and communication, and as a result knowledge is becoming deeper and also more complicated. Epistemic communities are growing rapidly in the developed countries. There is an increase in the epistemic communities from developing countries, too, but many of these are leaving their countries for foreign universities resulting in brain drain for developing countries and brain gain for developed countries. Thus, their expertise is often not available to their own countries. Besides, most developing countries have not institutionalized communication systems between the epistemic communities and the national policymakers and negotiators. This remains a serious problem (Gupta 1997; Gupta 2001b; Kandlikar and Sagar 1999).

(2) Rapid evolution in managed and spontaneous governance. As argued above there is significant evidence that governance approaches are being developed in two separate arenas—that of managed governance and that of spontaneous governance. Self-regulatory and regulatory approaches being implicitly and explicitly developed in the arena of spontaneous governance may have wide-reaching effects either because of their influence in highly integrated vertical markets or because of the way they create precedents that have a wide influence, or because they create confusion. Neither arena can be ignored by developing countries.

(3) Different governance options and five possible outcomes. As outlined before there are different clusters of initiatives being taken all of which are presently competing in the international arena. It is not adequate for developing countries to develop a strategy in support of some options and to focus only on those options. This might lead the developed countries to engage in forum shopping. Instead, developing countries have to develop a consistent strategy and push it forward in all competing fora through all relevant actors.

(4) Changing power politics. While tripartite—East, West, and Non-Aligned—politics was common in the period up to the 1990s, in the 1990s there was essentially North-South bargaining on global environmental and developmental issues. Since George W. Bush has taken over as president of

the United States in 2001, global politics have changed again and a new tripartite relationship is developing in global negotiations. This is the North-minus-the-US, the South, and the USA. The European Union is inching incrementally forward in environmental negotiations, while the United States is distancing itself from UN processes either through reservations (e.g. WSSD) or through non-ratification (e.g. the Kyoto Protocol, the Biodiversity Convention). Whether this is a short-term interruption of US policy, is unclear. What is clear, however, is that United States policy in the area of environmental treaties has been fairly consistent in the last twenty years, as the United States have not ratified several environmental treaties, including some pre-dating the 1990s, such as the UN Convention on the Law of the Sea and the Basel Convention. The few treaties that they have ratified—e.g. CITES, and the Vienna Convention and the Montreal Protocol—did not immediately threaten their domestic interests. In this changing context, the power of the developing countries is decreasing.

(5) The likelihood of a leadership vacuum. As argued before, another trend is that the average economic situation of the developing countries is deteriorating, even if at the margin there are countries that are doing better. Those that are doing better—such as China—are not evincing any desire to provide leadership and represent the cause of the developing countries in the international negotiations, as did the oil exporting countries in the NIEO discussions or Indonesia and India in the heyday of the Non-Aligned Movement. Although leadership, both intellectual and political is needed for developing countries, at this juncture, it looks as if in the coming decade this may not be forthcoming. Developing countries may have to learn to cope without leadership.

(6) Speed of the managed and spontaneous negotiating processes. All these governance initiatives are developing with rapidity. The formal negotiating process has been accelerating over the last three decades. While earlier treaties used to have conferences of the parties once in a few years (e.g. CITES), recent treaties (e.g. UNFCCC) have annual meetings with numbers of preparatory meetings each year. More regular reporting requirements are being created and more bodies to collect, collate and analyze the available science, the country reports and other documents are

being established, new instruments are being developed. Developing countries will have to learn to cope with this speed.

(7) Rules of procedure in disarray. The international rules of procedure that apply mostly to inter-state negotiation (managed governance) help to provide some degree of fair play in the international negotiation process. However, it seems useful to make a few points. First, these rules do not apply to spontaneous negotiating processes. Second, these rules are becoming extremely difficult to apply in international environmental negotiations. This is because although formally most negotiations only have two plenary sessions where the rules of procedure apply, in fact, most negotiations take place in a large number of simultaneously held working groups where translations, for instance, are not available. Most often developing countries do not have enough representatives to send to these different fora. Third, with increased empowerment of non-state actors in international negotiating sessions, the international negotiations are becoming flooded with new participants and information from these participants. At least in the climate change negotiations it is not clear if such participation actually increases the transparency of these negotiations or rather makes the process much more complicated, bureaucratic and confusing.[7] Fourth, the new instruments being developed in the different arenas may fall under different legal systems and it remains unclear what family of rules—i.e. private or public international law or something in between—is likely to apply.

(8) The development of principles and precedents. In the meantime the rapid negotiations within both the managed and spontaneous governance arenas are leading to the development of principles and instruments which are setting precedents for future negotiations that may be very difficult to reverse or change. Late-comers will have to adjust to these processes in the future. Regulatory competition implies that the best-prepared countries will

---

[7]   For a detailed analysis of this point see Gupta 2003b.

negotiate hard to get their ideas on record in the global system and the rest of the world will have to follow these rules.

Recommendations for Developing Countries

Many developing countries are still in the phase where they think that global governance is only a matter of diplomatic participation in international negotiations. They hang on to the notion of sovereignty as a shield to protect themselves. This is to overlook the fact that with the opening of the borders and globalization, governance issues can impact on them at all administrative and social levels. Sovereignty is no more than a conceptual shield. What is clear is that the international system is not going to slow down for the developing countries. The question is whether developing countries can continue with a business-as-usual attitude towards international environmental and developmental negotiations? The answer is that such a strategy is a risky one for these countries: 'The South must not remain a passive bystander in this process of change' (Nyerere et al. 1990, 271).

The South Commission had recommended in 1990 that in terms of substantive approaches, the developing countries needed to focus on development first, ensuring the effectiveness of development, overcoming the knowledge gap, protection of the environment and management of global interdependence (Nyerere et al. 1990). Having closely monitored the climate change negotiations myself and read the literature in other areas of negotiation, I argue that these five need to be considerably modified to address the management of global interdependence from a Southern viewpoint.

The Management of Global Interdependence

The management of global interdependence from a Southern perspective involves:

Articulating sustainable development. I would argue that an uncritical focus on development would merely reproduce the problems of the developed world at a global scale (Chatterjee and Finger 1994). Further, defining sustainable development as a goal is an unwieldy tool because in a rapidly changing world, it is extremely difficult to define such an end-point; instead such a discussion tends to make the whole debate extremely fuzzy. It would make more sense to focus on a Gandhian approach—define sustainable development as a process and the ends will take care of themselves. In such a process certain values such as sustainable use, good governance, public participation, interrelationship and integration, common but differentiated responsibilities, precautionary principle, etc. need to be protected. It is important that developing countries individually and jointly try and define these procedural values through a national dialogue process so that there is societal commitment to this.

Strategic approach to bridging the knowledge gap. In general, the knowledge gap is becoming bigger between the developed and the developing countries at least in the area of technology development.[8] A strategic approach to bridging this gap is needed. This calls for identification of where the knowledge gap is critical in that it affects the ability of the developing countries to (a) sell their surplus in the international market in the future (e.g. in the area of biotechnology and intellectual property rights, Shiva 1993, 2001) through the development of new rules in the international arena; (b) understand how environmental issues might have negative and/positive impacts on developing country livelihoods in the future, and (c) areas in which there is likelihood of dependence on foreign science and technologies in order to be able to function in the 21st century. Brain drain is also a major problem for these countries. This calls further for building strategic institutional channels to access the information in an efficient manner. A possible manner could be to organize e-conferences with expatriate and local scientists, NGOs and business actors to participate in sessions to collect and collate relevant information for negotiating processes.

---

8   This is not to deny that in the area of soft ware development, some developing countries are making tremendous progress; but there still remain large areas in which technological progress is limited.

Identifying the fora for intervention. Developing country governments have to recognize that three types of governance fora—horizontal, diagonal and parallel—exist in two arenas and that in order to be effective they have to be able to influence all fora simultaneously. This means that it is not enough for them to focus on being active in the inter-state processes to which they are invited, but to ensure that they are (a) aware of, and try to influence, other inter-state processes to which they are not invited, but which may affect them in the future (e.g. the now defunct Multilateral Agreement on Investments regime); (b) that they have a clear strategy in relation to how they will deal with diagonal agreements, some of which may involve provinces and/or municipalities and that they develop comprehensive guidelines for participation in these; and (c) that they identify which parallel initiatives are likely to influence them and develop a strategic approach to dealing with these.

Coping with the new challenges at vertical level. Governance today implies a number of up and down links within and outside the bureaucratic framework. This means that it is very inadequate if only the foreign affairs or environmental ministries at central level become involved. Implementation will have to be undertaken at domestic and even local level. The links downwards and upwards need to be actively reinforced in order to be able to generate ideas and solutions that are likely to have a high compliance-pull. Finally, they need to be aware of the consequences of scientific and bureaucratic networking on government and international policies and need to develop a strategic document about how to cope with these mostly invisible processes with strong effects.

Develop a process-related strategy. The development of a strategy on how to cope with the competing meetings in different fora in relation to governance issues would need to include: (a) a conscious strategy that aims at agenda building and the preparation of documents to support negotiations that set forth national positions, (b) the identification of national and international bodies that can help with the drafting of such text at short notice and that can undertake the responsibilities of scrutinizing national positions and drafting legal text, and (c) a conscious aim at building coalitions with like-minded countries to promote the national, regional and/or G-77 strategy. Nyerere et al. (1990: 211) conclude that self-reliance is the only functional strategy for the South in an unequal world. But while the strategy of self-reliance is necessary to create self-respect, it is an inadequate strategy in the face of globalization.

5   Conclusion

Let us now return to the question this chapter has posed. How are developing countries affected by the diverse options that become visible in global environmental governance? In order to address this question, this chapter has first argued that there are two arenas of governance—managed and spontaneous. In these two arenas, new forms of governance are visible representing new permutations and combinations of relationships within and between the state and non-state actor. It classifies the literature on governance as recommending nine options of governance. Not all of these are likely to have a promising future according to different schools of thought in the international relations literature. On the basis of a feasibility assessment, this chapter concludes that five scenarios of these options are likely to become dominant individually or jointly. These include multiple competing small regimes; possibly a situation where some functions or regimes may be clustered together to increase the efficiency, the possibility—though weak—of a vertically integrated strong centralized organization such as a world environment organization and the progressive development of international law. This chapter has then argued that on issues raised by the developed countries, developing countries face major negotiating problems which can be explained by eight mutually reinforcing theories on developing country negotiating strategy. As yet, developing countries have no focused strategy or policy in relation to parallel and diagonal governance, except to control them via processes of managed governance (e.g. control of eco-labelling schemes via the WTO negotiations).

In the context of globalization, the average developing country is faced by eight negotiation trends as identified in this chapter. In sum, this chapter concludes that a projection of existing trends would indicate a further marginalization of developing country concerns in the international governance arena. It further concludes that the momentum of the globalization process is such that it is not going to either slow down or make concessions to the developing countries in the negotiation process. This means that if the developing countries take themselves seriously and realize that they are gradually losing what little sovereignty they had to the multiple processes of globalization, they will have to develop a proactive strategy and prioritize an effective negotiating strategy at the international arena to cope with the new forms of horizontal, diagonal and parallel relationships based on a strong domestic vertical strategy. The core of this strategy is to develop one elaborate substantive and procedural policy which needs to be pushed for in all fora in both arenas simultaneously. In order to negotiate effectively, South-South trust is a pre-condition for any success in the global arena. This is something the developing countries need to work on seriously if they wish to be able to effectively represent their interests globally.

References

Agarwal, A., J. Carabias, M.K.K. Peng, A. Mascarenhas, T. Mkandawire, A. Soto, E. Witoelar (1992), *For Earth's Sake: A Report from the Commission on Developing Countries and Global Change*, International Development Research Centre, Ottawa.

Agarwal, A., S. Narain and A. Sharma (1999), *Green Politics: Global Environmental Negotiations*, Centre for Science and Environment, New Delhi.

Allott, P. (1999), 'The Concept of International Law', *European Journal of International Law*, vol. 10, 31–50.

Allott, P. (2001), 'Intergovernmental Societies and the Idea of Constitutionalism', in J.-M. Coicaud and V. Heiskanen (eds), *The Legitimacy of International Organizations*, United Nations University Press, Tokyo, 69–103.

Arend, A.C. (1996), 'Towards an Understanding of International Legal Rules', in R.J. Beck, A.C. Arend and R.D. Vander Lugt (eds), *International Rules: Approaches from International Law and International Relations*, Oxford University Press, Oxford, 289–310.

Bendell, J. (ed.), (2000), *Terms of Endearment: Business, NGOs and Sustainable Development*, New Academy of Business, London.

Bharucha, V. (1997), 'The Impact of Environmental Standards and Regulations Set in Foreign Markets on India's Exports', in V.G. Jha, G. Hewison and M. Underhill (eds), *Trade, Environment and Sustainable Development. A South Asian Perspective*, Macmillan Press, Hampshire and London, 123–41.

Biermann, F. and U.E. Simonis (1998), *A World Environment and Development Organization*, Development and Peace Foundation, Bonn.

Biermann, F. (2002), 'Strengthening Green Global Governance in a Disparate World Society: Would a World Environment Organisation Benefit the South?', *International Environmental Agreements: Politics, Law and Economics*, vol. 2 (4), 297–315.

Biermann, F. (2005), 'The Rationale for a World Environment Organization', in F. Biermann and S. Bauer (eds), *A World Environment Organization. Solution or Threat for Effective International Environmental Governance?*, Ashgate, Aldershot, 117–144.

Börzel, T. and J. Gupta (2000), 'A New North-South Conflict? Regulatory Competition in European and International Environmental Politics', Paper presented at European Concerted Action on the Effective Implementation of Environmental Law, Barcelona, 9–11 November.

Campins-Eritja, M. and J. Gupta (2002), 'Non-State actors and Sustainability Labelling Schemes: Implications for International Law', Non-State Actors and International Law, vol. 2 (3), 213–49.

Charnovitz, S. (2005), 'Toward a World Environment Organization: Reflections upon a Vital Debate', in F. Biermann and S. Bauer (eds), *A World Environment Organization. Solution or Threat for Effective International Environmental Governance?*, Ashgate, Aldershot, 87–115.

Chatterjee, P. and M. Finger (1994), *The Earth Brokers*, Routledge, London and New York.

Esty, D.C. (1994), 'The Case for a Global Environmental Organisation', in P.B. Kenen (ed.), *Managing the World Economy. Fifty Years After Bretton Woods*, Institute for International Economics, Washington DC, 287–309.

Gibson, R., (1999) (ed.), *Voluntary Initiatives: The New Politics of Corporate Greening*, Broadview Press, Peterborough.

Gosovic, B. (1992), *The Quest for World Environmental Cooperation: The Case of the UN Global Environment Monitoring System*, Routledge, London.

Grieco, J.M. (1996), 'Anarchy and the Limits of Cooperation: A Realist Critique of the Newest Liberal Institutionalism', in R.J. Beck, A.C. Arend, R.D. Vander Lugt (eds), *International Rules: Approaches from International Law and International Relations*, Oxford University Press, New York.

Gupta, J. (1995), 'The Global Environment Facility in its North-South Context', *Environmental Politics*, vol. 4 (1), 19–43.

Gupta, J. (1997), *The Climate Change Convention and Developing Countries—From Conflict to Consensus?*, Kluwer Academic Publishers, Dordrecht.

Gupta, J. (1999), 'North-South Aspects of the Climate Change Issue: Towards a Constructive Negotiating Package for Developing Countries', *Review of European Community and International Environmental Law*, vol. 8 (2), 198–208.

Gupta, J. (2000a), 'North-South Aspects of the Climate Change Issue: Towards a Negotiating Theory and Strategy for Developing Countries', *International Journal of Sustainable Development*, vol. 3 (2), 115–35.

Gupta, J. (2000b), *Climate Change: Regime Development and Treaty Implementation in the Context of Unequal Power Relations, Volume I: Monograph*, IVM-O-OO/02, Amsterdam.

Gupta, J. (2000c), *On Behalf of My Delegation: A Guide for Developing Country Climate Negotiators*, Center for Sustainable Development of the Americas, Washington DC.

Gupta, J. (2001a), 'Legitimacy in the Real World: A Case Study of the Developing Countries, Non-Governmental Organisations and Climate Change', in J.-M. Coicaud and V. Heiskanen (eds), The Legitimacy of International Organizations, United Nations University Press, Tokyo, 482–518.

Gupta, J. (2001b), 'Effectiveness of Air Pollution Treaties: The Role of Knowledge, Power and Participation', in M. Hisschemöller, J. Ravetz, R. Hoppe and W. Dunn (eds), *Knowledge, Power and Participation*, Policy Studies Annual, Transaction Publishers, New Brunswick and London, 145–74.

Gupta, J. (2002a), 'Global Sustainable Development Governance: Institutional Challenges from a Theoretical Perspective', *International Environmental Agreements: Politics, Law and Economics*, vol. 2 (4), 361–88.

Gupta, J. (2002b), 'The Climate Convention: Can A Divided World Unite?' in J. Briden and T. E. Downing (eds), *Managing the Earth: The Eleventh Linacre Lectures*, Oxford University Press, Oxford, 129–56.

Gupta, J. (2003a). 'Engaging Developing Countries in Climate Change: (KISS and Make-Up!)', in D. Michel (ed.), *Climate Policy for the 21st Century: Meeting the Long-Term Challenge of Global Warming*, Centre for Transatlantic Relations, Johns Hopkins University, Washington DC, 233–64.

Gupta, J. (2003b), 'The Role of Non-State Actors in International Environmental Affairs', *Zeitschrift für ausländisches öffentliches Recht und Völkerrecht / Heidelberg Journal of International Law*, vol. 63 (2), 459–86.

Gupta, J. (2004), 'Regulatory Competition and Developing Countries and the Challenge for Compliance-Push and Pull Measures', in G. Winter (ed.) *Institutional Dimensions of Earth Systems Analysis*, Transnational Institutions for the Environment, forthcoming.

Héritier, A., C. Knill and S. Mingers (1996), *Ringing the Changes in Europe. Regulatory Competition and the Redefinition of the State: Britain, France, Germany*, De Gruyter, Berlin, New York.

Hisschemöller, M. and J. Gupta (1999), 'Problem-solving through International Environmental Agreements: The Issue of Regime Effectiveness', *International Political Science Review*, vol. 20 (2), 153–76.

IDGEC (1999), *Science Plan, Institutional Dimensions of Global Environmental Change*, International Human Dimensions Report No. 9, Bonn.

ILA (2002), *The New Delhi Declaration of Principles of International Law Relating to Sustainable Development*, Resolution 3/2002 of the International Law Association, London.

Jha, V.G. (1997), 'Conclusions and Policy Recommendations', in V.G. Jha, G. Hewison and M. Underhill (eds), *Trade, Environment and Sustainable Development. A South Asian Perspective*, Macmillan Press Ltd., Hampshire and London.

Kandlikar, M. and A. Sagar (1999), 'Climate Change Research and Analysis in India: An Integrated Assessment of a North-South Divide', *Global Environmental Change. Human and Policy Dimensions*, vol. 9, 119–38.

Keohane, R.O., P.M. Haas and M.A. Levy (1993), 'The Effectiveness of International Environmental Institutions', in P.M. Haas, R.O. Keohane and M.A. Levy (eds), *Institutions for the Earth*, MIT Press, Cambridge (Mass.), 3–24.

Keohane, R.O. (1996), 'International Institutions: Two Approaches', in R.J. Beck, A.C. Arend, R.D. Vander Lugt (eds), *International Rules: Approaches from International Law and International Relations*, Oxford University Press, New York, 187–205.

Kersbergen, C.J. van and F. van Waarden (2001), *Shifts in Governance: Problems of Legitimacy and Accountability*, Social Science Research Council, The Hague.

Mannheim, J. (1999), *Corporate Conduct Unbecoming: Codes of Conduct and Anti-Corporate Strategy*, Tred Avon Institute Press, St. Michaels, Maryland.

Nath, K. (1997), 'Trade, Environment and Sustainable Development', in V. Jha, G. Hewison and M Underhill (eds), *Trade, Environment and Sustainable Development: A South Asian Perspective*, Macmillan Press Ltd., Hampshire and London, 15–20.

Nyerere, J. et al. (1990), *The Challenge to the South: The Report of the South Commission*, Oxford University Press, Oxford.

Oberthür, S. (2002), 'Clustering of Multilateral Environmental Agreements: Potentials and Limitations', *International Environmental Agreements: Politics, Law and Economics*, vol. 2 (4), 317–40.

Ramphal, S.S. (1983), 'South-South: Parameters and Pre-conditions', in A. Gauhar (ed.), *South-South Strategy*, Third World Foundation, London, 17–23.

Reed, D. (1993), 'Evolution of the Global Environment Facility', in D. Reed (ed.), *The Global Environment Facility: Sharing Responsibility for the Biosphere*, WWF, Gland, 3–8.

Schrijver, N. and F. Weiss (2002), 'Editorial', *International Environmental Agreements: Politics, Law and Economics*, vol. 2 (3), 105–08.

Schrijver, N. (1995), *Sovereignty over Natural Resources: Balancing Rights and Duties in an Interdependent World*, thesis, Rijksuniversiteit Groningen, Groningen.

Shiva, V. (1993), *Monocultures of the Mind*, Zed Books, London.

Shiva, V. (2001), *Protect or Plunder*, Zed Books, London.

Snyder, F. (2001), 'Global Economic Networks and Global Legal Pluralism', in G.A. Bermann, M. Herdegen and P.L. Lindseth (eds), *Transatlantic Regulatory Cooperation: Legal Problems and Political Prospects*, Oxford University Press, Oxford, 99–18.

South Centre (1993), 'An Overview and Summary of the Report of the South Commission', in South Centre (ed.), *Facing the Challenge: Responses to the Report of the South Commission*, Zed Books, London, 3–52.

Sridharan, K. (1998), 'G-15 and South-South Cooperation: Promise and Performance', *Third World Quarterly*, vol. 19 (3), 357–73.

von Moltke, K. (2002), 'Governments and International Civil Society in Sustainable Development: A Framework', *International Environmental Agreements: Politics, Law and Economics*, vol. 2 (4), 341–59.

von Moltke, K. (2005), 'Clustering International Environmental Agreements as an Alternative to a World Environment Organization', in F. Biermann and S. Bauer (eds), *A World Environment Organization. Solution or Threat for Effective International Environmental Governance?*, Ashgate, Aldershot, 175–204.

WSSD—World Summit on Sustainable Development (2002), *Johannesburg Declaration on Sustainable Development and the Plan of Implementation. Report of the World Summit on Sustainable Development, 2002*, UN Doc. A/CONF./199/20.

Young, O. (2002), *The Institutional Dimensions of Environmental Change*, MIT Press, Cambridge (Mass.).

Part II

The Case for a World Environment Organization

Chapter 4

Toward a World Environment Organization: Reflections upon a Vital Debate

Steve Charnovitz*

1   Introduction

Over a dozen years have elapsed since the run-up to the United Nations Conference on Environment and Development (held in Rio de Janeiro) when the idea of a World Environment Organization (WEO) began to receive serious attention. Although a spirited and illuminating debate has transpired, no real progress toward a WEO has occurred. Today, our planet still lacks effective global environmental governance.

During the same period, the world trading system succeeded in rationalizing and strengthening its institutional foundation. At the time of the 1992 Rio Conference, the General Agreement on Tariffs and Trade was in the sixth year of the multilateral Uruguay Round of negotiations, and prospects for a successful conclusion were in doubt. Some plans for establishing a new international organization were on the drawing board, but whether governments would agree to such a treaty and be able to ratify it in national parliaments remained a big question. The idea that the negotiations would lead to a powerful, rule-based organization with a

* The author wishes to thank Christopher D. Stone and Stephanie St. Pierre for helpful comments.

judicial system that can expound law, and that membership in the new entity would increase from 110 to 147 countries after nine years, was hardly the consensus scenario at that time. Clearly, governance for trade has enjoyed a much better recent run than governance for environment.

Today, the World Trade Organization (WTO) is in the midst of another negotiating round originally set to conclude in 2005. Efforts are underway to strengthen and broaden trade rules. Several environmental issues are on the negotiating table in Geneva, including reducing trade barriers to environmental goods and services and controlling fishery subsidies. Using comparative institutional analysis, commentators have asked why the WTO appears to be so effective at accomplishing its mission, and at expanding its mission, while the environment regime seems relatively less capable. One answer is that the trade regime is centred in an effective international organization while the environment regime is not (German Advisory Council on Global Change 2001, 176-77; Ostry 2001, 290-93).

Since the early 1990s, many analysts have called for correcting this organizational dysfunction in the environment regime. The opening salvo in the contemporary debate[1] was the perceptive article by Sir Geoffrey Palmer, the former Prime Minister of New Zealand, who examined 'New Ways to Make International Environmental Law' (Palmer 1992). Sir Geoffrey called for the establishment of an 'International Environment Organization' within the UN system. In 1994, Daniel Esty began an intellectual campaign for a Global Environment Organization (GEO) that would develop a comprehensive international response to environmental challenges (Esty 1994, 230).[2] Several other analysts have also advocated a

---

[1] The idea of an international organization for the environment certainly predates 1992, but this article focuses on the period since then. The quest for better international institutions for the environment began in 1909 with separate efforts by Theodore Roosevelt and Paul Sarasin. A recent volume by Martin Holdgate discusses this early history briefly, and has good chapters on the progress made transgovernmentally in the late 1940s (Holdgate 1999). The next renaissance for environmental governance was the early 1970s. One analyst of that era was Abram Chayes (see Chayes 1972).

[2] This volume uses the term World Environment Organization (WEO), and so that convention will be followed in this article. Nevertheless, a 'GEO' would be a much better name for such an institution. The term 'GEO' is also the name for the Global Environment Outlook prepared by the UN Environment Programme.

new environmental organization, such as Ford Runge (Runge 1994, 100; 2001), Frank Biermann (Biermann 2000; 2001), the Zedillo Commission (High-Level Panel on Financing for Development 2001, 26), Gus Speth (Speth 2002b, 22-23), the Shadow G-8 group (Shadow G-8 2003, 29), and the team of John Whalley and Ben Zissimos (Whalley and Zissimos 2001; 2002). Some thoughtful analyses of the various WEO proposals have now appeared (Hierlmeier 2002; Lodefalk and Whalley 2002; Marshall 2002; Gaines 2003; Haas 2004; Oberthür and Gehring, this volume).

Two responses have emerged concerning these ideas for more cohesive environmental governance. Several environmental analysts have criticized these proposals for being over-ambitious, centralist, pro-North, unsophisticated, or unnecessary (for example, see Juma 2000; Najam 2002 and this volume). Often, however, the WEO proposals have been overlooked. For example, a major new study of environmental governance omits any attention to the debate regarding the WEO (UNDP et al. 2003).

The concept of a WEO was only briefly mentioned in the preparatory sessions for the 2002 World Summit on Sustainable Development (WSSD), and was not discussed at all at that Johannesburg Summit. In 1997, at the Special Session of the UN General Assembly, four governments had proposed consideration of a 'global environmental umbrella organization of the UN', and there were expectations that such ideas would be advanced over the following five years. Instead, the UN General Assembly took a side-step by convening an annual Global Ministerial Environmental Forum (GMEF), which was set up merely as a periodic 'Forum' rather than a continuing organization. When the GMEF held its first meeting, at Malmö in 2000, the Forum called for the forthcoming 2002 conference (later termed the WSSD) to 'review the requirements for a greatly strengthened institutional structure for international environmental governance based on an assessment of future needs for an institutional architecture that has the capacity to effectively address wide-ranging environmental threats in a globalizing world' (Malmö Declaration 2000, para. 24). At the same time, the UN Environment Programme (UNEP) launched a series of meetings on international environmental governance. This series concluded in Cartagena in early

2002 with recommendations for a stronger funding base for UNEP, and for the GMEF and UNEP Governing Council to be utilized more effectively.[3] Not surprisingly, the UNEP-sponsored meetings did not recommend that UNEP be replaced by a WEO. By the time of the Johannesburg Summit, the construction of a WEO was no longer being actively considered.[4]

Notwithstanding this reticence at the WSSD about environmental governance, the delegates were eager to pontificate on trade governance. For example, the Johannesburg Plan of Implementation (2002, paras. 47, 90, 96) encourages efforts by international financial and trade institutions to have more open and transparent decision-making processes, urges WTO members to facilitate the accession to the WTO by developing countries, and calls for action at all levels to enhance trade infrastructure and strengthen institutions. To be sure, the amount of text in the Plan on Implementation regarding trade is much less than the amount regarding environment. Still, it seems noteworthy that the governments were willing to delve into the management of the trading system while not paying much attention to needed improvements in environmental governance.

Two possible explanations exist for why the results from the WSSD were so meagre on governance, and for that matter on environmental stewardship in general. The first is that the environment regime is running smoothly and the other is that it is so poorly designed that it cannot reform itself incrementally. Explanation No. 1 is obviously wrong: Our planet faces significant environmental challenges (Turner 2000; Worldwatch 2003; Speth 2004), and I am not aware of any serious analyst who claims that current governance of the Earth's 'ecolonomy' is sufficient. The second explanation is that the environment regime has a vested interest in maintaining its compartmentalization, and will strongly resist any consolidation. If this second explanation is right, then the prospects for reform are dire indeed.

On the assumption that environmental governance needs fixing, what is to be done? Proponents of a WEO should reflect on why so little has

---

3   See http://www.iisd.ca/unepgc/gmef3/.
4   The UN University held a panel session on governance at the third WSSD PrepCom, and the papers are now in publication (see Chambers and Green 2004).

been accomplished toward that goal. One problem is that the advocacy for a WEO has not been convincing. Anyone who examines the various proposals would see an air of first positing reorganization and then searching for a mission. Proponents of a WEO will need to renew efforts to make a more cogent case for reform.

The frustration with the stalled debate was undoubtedly a reason why Esty has shifted to advocating a more gradualist 'Global Environmental Mechanism' that would link together existing institutions and add new structures when warranted to carry out core functions of environmental governance. In a recent paper, Esty and his colleague Maria H. Ivanova point to the need for better data collection, compliance monitoring, scientific assessment, bargaining, rule-making, civil society participation, financing, technology transfer, dispute settlement, and implementation coordination mechanisms (Esty and Ivanova 2002).

The Johannesburg Summit presented an opportunity to address these gaps within existing governance structures. Unfortunately, the governments did not do so, and also failed to consider the organizational prerequisites for environmental policy.

The newest model for a WEO is the proposal from Whalley and Zissimos for an organization to help governments and private actors do environmental bargaining. This initiative, funded by the MacArthur Foundation, is creative and useful. But at best it could deliver only a partial solution to current governance problems. Building on Ronald Coase's classic analysis of how polluters and victims could bargain to achieve joint gains so long as the liability rule is clear and transaction costs are low, Whalley and Zissimos extend the argument beyond pollution into the use of natural resources. They suggest that a global mechanism could lower transaction costs and facilitate negotiated exchanges, and perhaps help to clarify property rights. The ultimate goal is to achieve full internalization of cross-border externalities so that those who undertake economic activities bear their full environmental costs.

Whalley and Zissimos are right that considerable scope exists for international bargaining beyond the quantity of deal-making occurring now. This gap certainly suggests the need for better mechanisms to assure contractual performance in international environmental deals (see Stone 1993, 39-42). But Whalley and Zissimos go further than that in boldly arguing that other WEO proposals are not focused on central and substantive environmental policy problems, and that their own plan could achieve more than the current network of issue-specific environmental treaties (Whalley and Zissimos 2002, 164-66).

In my view, the Whalley and Zissimos proposal to enhance markets is worth trying, but it is hardly a sufficient solution to the challenges of environmental protection. The subtext of their proposal is that the rich countries will pay poor countries for guaranteed outcomes such as preserving a rainforest. Although a higher volume of exchanges might be possible, one wonders how deep the pockets are in rich countries for such monetary deals. While no one can deny the potential benefit of proper pricing for environmental resources, Whalley and Zissimos do not offer any reason to believe that bargaining will occur on a sufficient scale to achieve a significant amount of cost internalization. In one revealing passage, the authors note that their proposal focuses on 'cross-border externalities since within-country externalities can in principle be dealt with by solely domestic initiatives' (Whalley and Zissimos 2002, 166). Yet their study provides no evidence that robust domestic markets for such bargaining currently exist and are achieving significant cost-internalization. If such bargaining does not actually occur domestically, why imagine that it will occur across borders? Perhaps what Whalley and Zissimos meant by 'domestic initiatives' is government-imposed taxes and regulations. Yet if such a regulatory strategy is needed *within* each country, then why would one pursue an entirely different strategy for transborder issues as a substitute for regulation through treaties and, when justified, extraterritorial application of law? The next stage of the Whalley and Zissimos project should consider these points.

The purpose of this article is to restate the case for a WEO. The remaining discussion has three parts: Section 2 will seek to explain why a WEO is needed by examining, in turn, the W, the E, and the O. Section 3 will suggest that the paradigm for the WEO should be *competition*, as well as cooperation, the goal stressed in the pro-WEO literature. In both of these parts, the article will take note of the WTO, and point out where it is a good model for a WEO or a poor model. The article ends with a short conclusion (section 4).

2   Why a W-E-O is Needed

Because ecosystems overlap political units, it stands to reason that international and/or transgovernmental organizations will be needed to manage human interaction with the environment. States alone may be able to perform this function with respect to environmental problems residing within national borders (e.g., noise pollution), but most of the serious environmental problems today are transborder and/or global. For those, solutions will require collective action. Truly effective international environmental institutions can improve the quality of the global environment (Keohane, Haas, and Levy 1993, 7). Institutions help by increasing governmental concern, by building capacity, and by enhancing the ability to make and keep agreements (Levy, Keohane, and Haas 1993, 398, 424).

Externalities occur when a producer or consumer does not take into account the adverse effects that it imposes on others. Such market failure is the core problem of environmental policy. Responses to this problem include regulation, clarification of property rights, and facilitating bargains. As André Dua and Dan Esty have pointed out, when externalities traverse national borders, they can be viewed as 'super externalities' because of the additional hurdle they present of securing cooperation among sovereigns (Dua and Esty 1997, 59).

Governments began responding to transborder environmental problems in the 19th century through treaties, and, during the 20th century, drafted hundreds of new treaties and established scores of international organizations with responsibility for environment, natural resources, and global public goods. The record shows that governments have been willing to initiate cooperation on specific problems by establishing conventions and related institutions. Yet few governments have shown a willingness to meld these institutions into a holistic entity.

Although UNEP, established in late 1972, has helped to promote international environmental law (Tolba 1998; McNeill 2000, 350), the organization has chronically underperformed. The problem is not quality of leadership. Over 30 years, UNEP's executive leadership has been better than average for international organizations, and its current Executive Director, Klaus Töpfer, is quite capable. The problems of UNEP stem from its low status within the UN bureaucracy, its disadvantageous and dangerous location in Nairobi, its inadequate and insecure funding, and its detachment from many of the multilateral environmental agreements (e.g., climate change).

The 'trade and environment' debate of the 1990s stimulated many outsiders to examine UNEP, and that Programme looked feeble organizationally in comparison to the trading system. UNEP's sorry state, including its weak presence and staffing in Geneva, triggered recommendations for a world organization for the environment that could operate in equipoise with the WTO. For example, Supachai Panitchpakdi, now the Director-General of the WTO, once stated that 'the problem is that there is nobody of the same stature to deal with the WTO because there is no World Environment Organization' (Supachai 2001, 443). The institutional strength of the WTO reflects an acknowledgement by governments that economic interdependence is a fact, and that nations will be better off with a robust organization that can help manage that interdependence. Yet ironically, even though the global environment is more of an integrated system than the global economy (Speth 2002a, 13), governments have not drawn a parallel conclusion about the value of a WEO.

Why A World Organization?

A common complaint about creating a WEO is that such an organization would be too powerful and intrusive. In that respect, the WTO analogy has hurt the pro-WEO cause because of the political baggage the WTO now carries. The developing countries, as a generalization, view the WTO's rule-based approach as being too coercive to them, and too insensitive to national development plans. Some groups in civic society view the WTO as the champion of harmful 'globalization from above' because it promotes economic integration and elevates decision-making to a level beyond the influence of the public. Another complaint about the WTO is that even though each governmental member ostensibly has the same influence, in practice richer countries have a greater say. Thus, advocates of a WEO now have the burden of explaining how a WEO will avoid these reputed problems of the WTO.

To the extent that a WEO would be a centralized, top-down institution, that seems to rub against the grain of 'subsidiarity' (Newell

2002, 669), a term from European law positing that authority to make decisions should not be raised to a higher level (the Community) when a lower level (a Member State) would be adequate (see Bermann 1994; Vergés 2002).

The term 'subsidiarity' originated in Catholic philosophy. In his 1931 Encyclical 'Reconstruction of the Social Order', Pope Pius XI explained that 'Just as it is gravely wrong to take from individuals what they can accomplish by their own initiative and industry and give it to the community, so also it is a grave evil and disturbance of right order to assign to a greater and higher association what lesser and subordinate organizations can do' (Pius XI 1931, para. 79). The Pope termed this principle the 'subsidiary function', and called on those in power to pursue a graduated order. Most of the discussion about subsidiarity in the Encyclical focuses on the State vis-à-vis the individual. The Pope did not discuss this principle with respect to governments in the League of Nations (or, for that matter, with respect to decision-making in the Catholic Church). Thus, one cannot assume that the doctrine will always apply to the relationship between an intergovernmental organization and its member States.

States may have valid reasons to delegate decision-making upward to international entities. Doing so may enhance the dignity of the individual even though decision-making may be slightly more remote. Although subsidiarity is sometimes characterized as a principle of non-interference, this shorthand misses the duality in the Pope's discussion which is as much concerned with helping smaller units as it is with not interfering with them. As a scholar of subsidiarity explains, each larger grouping is understood to serve the smaller, and all in the end are understood to serve individual dignity (Carozza 2003, 43).

In any event, a commitment to subsidiarity does not present a true stumbling block for WEO advocates, who point out that a WEO is only needed for those problems that are *not* being solved at a lower level. Numerous global problems exist, such as climate change, ozone depletion, ocean pollution, and fisheries depletion, and so there would be plenty of issues for which a WEO might be the right level to assign a lead competence. Ironically, none of those particular issues is now overseen by UNEP; all have been assigned to other organizations or treaty entities. Many WEO advocates would also assign it responsibility for helping governments address transboundary environmental problems (such as hazardous waste), and the maintenance of public goods (such as biodiversity). The subsidiarity rationale for allowing a WEO to share jurisdiction over such issues is that the lower organizations (i.e., the national governments) consent because they need coordination from above.

The case for a world-level response is probably the weakest for the *common* challenges that all countries face, such as clean water, waste disposal, etc. Yet it was precisely that genre of issues (rather than, say, climate change), that were the centrepiece of the WSSD in 2002. To my knowledge, no government raised an objection in Johannesburg to devoting so much time to issues that are inherently local. That focus at Johannesburg was similar to the orientation of the UN Millennium Development Goals. For example, one of the Goals is to reduce by half the proportion of people without sustainable access to safe drinking water.

Any international environmental entity is accountable with regard both to its outcomes and its procedures. Some analysts fear that a global organization cannot possibly appreciate the subtleties of environmental policy appropriate for separate communities around the world, and thus a WEO's norm-generation and other activities could wreak unintended harms. Although the establishment of the WTO enjoyed the support of international business groups, the establishment of a WEO has not drawn symmetric support from international environmental groups. Surely, one reason why is that many nongovernmental organizations (NGOs) fear that a WEO might make it harder for citizens and associations to influence environmental policy. The doubts about accountability are sometimes expressed as concerns about a 'democratic deficit' or a gap in 'legitimacy'.

Legitimacy has many facets. The least controversial claim is that an international organization should act legitimately with respect to its member governments. The WTO has a mixed record on that facet of legitimacy. On the one hand, it has rule-based legitimacy in that all actions putatively have to gain the consensus of all member governments. On the other hand, the traditional WTO practice of handling controversies by having key governments convene privately in a 'green room' has not yet ceased (Global Accountability Report 2003, 15). In the environment regime, there has been an effort over the past few years to reconstitute the UNEP Governing Council to include all governments based on the rationale that universal participation is more legitimate than representative participation (see Johannesburg Plan of Implementation 2002, para. 140d).

That rationale is questionable, however, if a larger assembly would make decision-making harder.

A more contested claim is that the constituents of an international organization include the public in each country, if not a global public.[5] The counter-argument is that international organizations are sufficiently accountable to each individual in a transitive fashion through his or her own government. Whether or not that is true as a matter of democratic theory, many individuals and groups (e.g., anti-globalization protestors) believe that the legitimacy of an international agency is undermined by the lack of a direct connection to an electorate. NGOs have also criticized the insularity and secretiveness of some international organizations, such as the WTO.

When the problems considered are complex and solutions emerge slowly—the common predicament on global environmental issues—the value of transparency and regular public input becomes obvious as a way of securing better information and ideas. The WTO achieves some transparency, but scores badly on eliciting public input. Thus, while a WEO might be able to copy the WTO approach for how governments participate, a WEO would need to be far more open to civic society and business than the WTO is. This would be in line with the Rio Declaration which states that 'environmental issues are best handled with the participation of all concerned citizens, at the relevant level' (Rio Declaration 1992, Principle 10).

Given the long-time practice of nongovernmental participation in environmental governance, there would seem to be little point in establishing a WEO if based on the common intergovernmental model in which NGOs participate as a sideshow. Indeed, the prospects for a broad-based WEO could enhance public support for adopting a new organization. If a WEO is to be created, it should reflect the learning from the path of the

---

5   It is interesting to note that the first paragraph of the Johannesburg Declaration on Sustainable Development begins 'We, the representatives of the peoples of the world . . .' (Johannesburg Declaration 2002). One sees a similar populist theme on the home page of the UN website which begins, 'United Nations. It's Your World'.

International Labour Organization (ILO), a body created in 1919. In the ILO, workers and employers participate equally with government representatives, a feature known as 'tripartism'.

ILO-style tripartism, however, is no longer fully adequate as a template for gaining NGO participation. An effective WEO would have to provide space for government representatives to work with a multiplex of stakeholders including environmental NGOs, human rights groups, businesses, scientists, religious leaders, mayors of cities, and many other stakeholders. Such inclusiveness would be in line with the doctrine of subsidiarity, which states that 'social activity ought of its very nature to furnish help to the members of the body social, and never destroy and absorb them' (Pius XI 1931, para. 79).

Should governments be unwilling to extend participation in a WEO to private social and economic actors, then that would tip the scale in favour of those who demur that the cost of a massive reorganization into a WEO would be too high relative to the expected gains. Why go to the trouble of setting up a new international organization if it is to be composed merely of government officials and bureaucrats? One does not need an international organization for governments to cooperate; they can do so bilaterally, or through emerging transgovernmental networks (see Raustiala 2002). Yet on many international problems, governments will be too cautious and nationalistic to reach integrative solutions without the catalyst of nongovernmental input.

Establishing a participatory WEO would be a challenge, and no recipe for it exists. Perhaps the most difficult task is to find a way to combine broad participation with a decision-making capacity for the organization.[6] The sorry experience with the UN Commission on Sustainable Development (CSD) (see Elliott, this volume) stands as a stark

---

6   The ILO achieves this. It is true that ILO Conventions are not law until they are ratified by governments. But a requirement for approval of new law at the national level is the norm for all specialized organizations, including the WTO and the World Health Organization. The only major international organization with authority to write new rules that are automatically obligatory is the UN Security Council.

reminder of the pointlessness of fostering broad participation detached from any decision-making responsibility.

Sometimes, analysts argue that the failures in international environment policy are not caused by poor organization, but rather are caused by lack of political will (see Najam 2002, 8). Yet that diagnosis seems to miss the point that well-designed international institutions can help generate political will by constructing new social norms.

Environmental Protection and Sustainable Development

Sustainable development is a useful concept (Tarlock 2001; Holliday, Schmidheiny and Watts 2002). It marries two important insights: that economic development should be ecologically viable and that environmental protection does not preclude development. Sustainable development also has value in positing an answer to the trade-off between the welfare of the people today and the welfare of the people of the future.[7] Thus, the goal of sustainability should inform the work of all international agencies (Dowdeswell and Charnovitz 1997, 101).

Yet it is one thing for 'sustainable development' to be an inspiration, and quite another for that vague term to be the organizing principle for governmental action. In the years since the Rio Conference of 1992, during which 'sustainable development' has reached mantra status,[8] we have not seen many examples of how the concept has made much of a policy difference (Esty 2001). It may be unfair to point to the CSD as an example of such failure because the CSD was not set up to do anything except be a talk shop. Yet that is exactly the point: when the international

---

[7] Sustainable development is commonly defined as development that meets the needs of the present without compromising the ability of future generations to meet their own needs. See http://www.un.org/esa/sustdev/about_us/aboutus.htm.

[8] For example, the ILO Declaration on Fundamental Principles and Rights at Work (1998) refers in preambular language to the goal of 'broad-based sustainable development'. See http://www.ilo.org/public/english/standards/decl/declaration/text/index.htm. See also the UN Millennium Goal no. 7 to 'Integrate the principles of sustainable development into country policies and programmes; reverse the loss of environmental resources', available at http://www.un.org/millenniumgoals/.

community glorifies a concept as expansive and ambiguous as sustainable development, perhaps the best place for it *is* a talk shop.

One unfortunate manifestation of the 'sustainable development' concept is that it is elbowing out environmental protection at the international level. Recall that in 1972, the world community held a UN Conference on the Human Environment (in Stockholm), and, in 1992, held the UN Conference on Environment and Development (MacDonald 2003, 166-68). In 2002, however, there was a decennial conference called the World Summit on Sustainable Development. Thus, over the years, we have seen the premier global environmental event transmogrify from a conference focused on the environment, to a conference about environment and development, and then to a Summit where the term 'environment' has been banished from the event's title. Ironically, the Johannesburg Conference was held just a few months after another major development conference, the UN Conference on Financing for Development (in Monterrey). That propinquity itself provided a reason to rehabilitate the Johannesburg Summit back to environment, but the governments did not even consider that. In recent years, UNEP, too, has shifted its attention more toward development, and has adopted a new motto, 'Environment for Development' (UNEP 2002, 4).

In stating that environmental problems should be dealt with directly, I do not mean to sound anti-development or anti-poverty reduction. Certainly, the plight of the poorest countries may be the central economic and moral issue of our time. In my view, the United Nations and other international organizations, such as the WTO, should be doing a lot more to alleviate poverty. But that does not justify usurping an environmental agenda with a poverty reduction agenda. Both agendas are important and distinguishable from one another. Trying to meld them ends up short-changing both. To be sure, defenders of the Johannesburg Summit would argue that poverty reduction can be a potent environmental strategy, and I agree. Yet that is hardly a reason to refrain from holding a Summit to zero in on environmental challenges.

If the Summit had succeeded in delivering significant environmental benefits, then the continued reliance of a paradigm of 'sustainable development' might be more supportable. The absence of such a positive result demonstrates the inefficacy of organizing global meetings around the sustainable development objective. Space does not permit a full auditing of the meagre output from Johannesburg, so I call attention to the document 'Key Outcomes of the Summit', located on the UN web-site.[9] The eight key outcomes listed can be summarized as: (1) reaffirming sustainable development, (2) broadening sustainable development to include poverty linkages, (3) issuing concrete commitments and targets for action, (4) giving attention to energy and sanitation issues, (5) supporting a world solidarity fund for eradication of poverty, (6) focusing on the development needs of Africa, (7) taking into account the views of major groups, and (8) boosting partnerships with the private sector. Of those, outcomes 2, 4, 5, and 6 are not really environmental. Outcome 2 and 7 are about process, and outcome 1 is regurgitive.

Thus, if there was any environmental policy advance in the Summit, it has to be in outcome number 3, the concrete commitments and targets. Yet according to the accompanying fact sheet regarding those commitments, many of the targets hew to development rather than environment (e.g., poverty eradication, sanitation, infant mortality, and energy), and some of the ones that are environment (e.g., safe drinking water) are just restatements of goals previously established by the United Nations. Boiling all this down, there are just a few new environmental targets—for chemicals (2020), water efficiency (2005), oceans (2010), fish stocks (2015), and biodiversity (2010)—yet even there, no specific goal is backed up with an action plan likely to achieve the goal. Given the many months of planning for Johannesburg and the numerous 'PrepComs' and planning sessions held, the wispy output of the Summit did not help environmental policy escape the doldrums of the past few years (see Gutman 2003).[10]

---

9 See http://www.un.org/partners/civil_society/calendar/c-dvcop.htm. Another self-congratulation on the UN website about the 'Implementation Summit' says that one of the 'major accomplishments' was 'strengthening of the concept of sustainable development and the important linkages between poverty, the environment and the use of natural resources'. See 'The Road from Johannesburg: What was achieved and the way forward', at http://www.johannesburgsummit.org/.

10 A few positive outcomes from Johannesburg are worth noting. One was the programmatic emphasis on initiating partnerships between governments, business, civil society, and international organizations. Another was the Global Judges Symposium which brought together senior judges from 59 countries and from international courts and tribunals. The

Were sustainable development a viable programmatic objective, one could expect to see—in the 12 years since the Rio Conference—numerous governments setting up ministries of sustainable development. While some do exist (e.g., France), that is hardly a common feature of national administration. Instead, governments have continued to maintain environmental ministries that are separate from trade ministries and energy ministries. Recognizing the separateness of environment and sustainable development objectives is certainly consistent with the recommendations of the Brundtland Commission which stated in 'Our Common Future' that 'Environmental protection *and* sustainable development must be an integral part of the mandates of all agencies of governments, of international organizations, and of major private-sector institutions' (World Commission on Environment and Development 1987, 312 emphasis added). In my view, most governments made the right choice in avoiding conglomeration through a sustainable development ministry because such a ministry would probably be ineffective.

Progress in national government and international governance over the years has come through specialization (Gardner 1974, 558). This functional approach is not controversial in most international bodies. One expects the ILO to bring together labour ministers to focus on workers and employment. One expects the WTO to bring together trade ministers to focus on trade. One expects the World Health Organization (WHO) to bring together health officials to focus on disease and public health (see Stein 2003, A15). By contrast, for the environment, when the United Nations holds a world conference, it is apparently not politically correct to bring together governments to focus on ecological threats. This skittishness has gotten worse in recent years, and is a main reason for the miasma in international environmental governance. To be sure, the meetings of the UNEP Governing Council and the GMEF do convene environment

judges adopted the Johannesburg Principles on the Role of Law and Sustainable Development (available in UNEP/GC.22/INF/14). Still another was the inauguration of the 'Partnership for Principle 10' to promote good environmental governance at the national level, including transparency, participation, and access to justice. Principle 10 was part of the Rio Declaration of 1992. The new Partnership includes governments, international organizations, and NGOs. See www.pp10.org.

ministers.[11] Such a meeting was held in February 2003 in Nairobi, but did not accomplish much (see Nanda 2003). Another meeting was held in March 2004 in Jeju, but whatever result it generated has not yet been released to the UNEP website.[12]

Despite the initially high expectations, the WSSD brought the worst of both worlds for the environmental regime. Although widely perceived as the once-in-a-decade opportunity for national leaders to address the environment, the Summit ended up being more about poverty (and then only rhetorically). With that political space now having been used up, the environment will probably not gain another Summit this decade. Any attempt to upgrade environmental governance will have to contend with the albatross of 'sustainable development', and demands by developing countries that any new organization have sustainable development as the core objective. So what is wrong with that: why not take 'sustainable development' seriously and organize internationally around that overarching concept? The reason why not is that governments are loathe to let an organization for sustainable development interfere with other functional international organizations.

This attitude was apparent at Johannesburg. Many developing countries, quite logically, took the position that a conference on sustainable development had the competence and responsibility to address international trade policy. These governments pushed for new commitments on trade liberalization, but ran into the buzzsaw of European and US delegates who did not want even to discuss going beyond what had been agreed to at the WTO one year earlier (Gray 2003, 258-65). As a result, although there is a great deal of verbiage in the Johannesburg Plan of Implementation about trade, all of it merely rephrases prior agreements that have been reached at the WTO.

---

[11]   This is generally true, but not in the United States which lacks an environmental ministry with international competence. The US Department of State represented the United States at the Johannesburg Summit, and also does so at the GMEF and the UNEP Governing Council.

[12]   As of 27 June 2004.

The same concern about turf exists in the World Bank, the International Monetary Fund, and other UN organizations. All would resist having a Sustainable Development Summit intrude on their policy mandates. Despite the ostensible allegiance to sustainable development, governments are not going to use a Johannesburg Summit or analogous future event to supervene the competence of other international agencies.

Yet if a 'Summit' about sustainable development cannot negotiate on trade, or development funding, or intellectual property, then it is a hollow Summit. The same point holds for a prospective World Organization for Environment and Development (for one proposal, see Simonis and Brühl 2002, 122–23). The necessary comprehensiveness would make it impossible to create such an Organization with a meaningful mandate.

In conclusion, if a WEO is to be set up, its mission should be to address the top environmental risks facing the planet. A WEO could also address the growing inconsistencies *between* environmental conventions, a problem that now lacks an organizational solution (see Wolfrum and Matz 2003).

Organization and its Discontents

This leads to the final consideration: Is the 'O' in WEO realistic? Critics of such an Organization have made two salient points. One is that incremental improvements in current governance will have to be adequate because nothing else is feasible. The other is that the environment regime is too complex for one WEO.

The notion that the current environment regime is the best that humans can accomplish would be a preposterous claim. At present, the regime is one of *disorganization* with hundreds of agencies and treaties operating unlinked to each other. Not once have governments taken the time to design an ideal management structure. Instead, whenever a new environmental problem arose, a new entity was opportunistically added. Rarely have entities been dismantled, even when they are so obviously ineffective, such as the CSD.

Another disappointing experience is with the G-7/G-8 Environment Ministers, who have been meeting annually for nine years without much to show for it.[13] At the most recent meeting in April 2003, the Ministers announced support for increased environmental coordination at the international level through broad policy guidance and advice of the UNEP Governing Council/GMEF and 'full respect for the independent legal status and governance structure of other entities ...' (G8 Environment Ministers Communiqué 2003). That G-8 meeting heard a proposal from France to consider establishing a new UN Organization for Environment, but there was apparently minimal support for the idea (Speer 2003).

The organizational failure in environmental governance is especially disturbing when one compares it to the more rationally-organized trade regime. An organigram of the WTO shows a political ministerial body, a hierarchy of policy committees, a dispute settlement system, and a group that reviews national government policies.[14] To my knowledge, no organization chart for the complete environment regime(s) even exists. If it did, it would be a mishmash, with numerous boxes unconnected to each other.[15] The costs of such organizational anomie are high.

Whether governments and stakeholders are saddled forever with this disorganization is a matter of conjecture. While I agree that establishing a WEO would be difficult politically, I cannot accept that it is the 'organization of the impossible' to use Konrad von Moltke's memorable phrase (see von Moltke 2001). Good environmental policy is no longer just a preoccupation of the rich countries; it is equally sought by new environmental leaders in developing countries too (French 2003).

Certainly, a WEO will not be set up unless there is a large group of governments and stakeholders who demand it. Unfortunately, we are far away from that. The major multilateral environmental agreements have spawned distinct epistemic networks that seem to have a vested interest in

---

[13]   For a more positive view, see Kirton, this volume.
[14]   See http://www.wto.org/english/thewto_e/whatis_e/tif_e/org2_e.htm.
[15]   By contrast, the new World Resources report glorifies the current structure as a 'Symphony of Organizations' (UN Development Programme et al. 2003, 139).

maintaining a highly compartmentalized system. The uneasiness among environmentalists about a WEO will have to be reversed before any progress can be made.

Although a claim that the current regime is sufficient is unsupportable, a strong argument does exist that the totality of environmental issues and international environmental entities is far too extensive to be immediately joined into one organization. Thus, any initial WEO will necessarily have to be far less than comprehensive. Based on this reality, a good first step toward reform might be to cluster related multilateral environmental agreements (MEAs) into three or four groups (see von Moltke, this volume), to build new environmental organizations around them, and then perhaps to abolish or redefine UNEP. Such a plan would emphasize the linkages among related treaties and environmental entities.

Nevertheless, clustering has its own pitfalls (see Biermann, this volume). Whatever clusters are designed will leave out important links between the cluster. Furthermore, all of the functional tasks identified by Esty and Ivanova (2002), such as data collection and monitoring, would seem to be a cross-cutting feature of each cluster. Similarly, the bargaining proposed by Whalley and Zissimos (2002) would be stunted if it had to occur within each cluster, rather than across clusters.

Alternatively, the first step could be to establish the WEO initially with only planning and budget functions. The WEO could seek to address the biggest flaw in the status quo, which is that no ongoing mechanism exists to identify the most serious gaps in the stewardship of environmental resources and to determine where new environmental investments are most needed. Such a WEO could hold annual conferences at the ministerial level and more frequent meetings on particular topics. It could also set up a inter-parliamentary assembly to serve a consultative role.

Of all the existing international environmental entities, the Global Environment Facility (GEF) is perhaps the best model for a more extensive environmental organization (Streck 2001). It focuses on six critical global environmental threats—biodiversity loss, climate change, degradation of international waters, ozone depletion, land degradation, and persistent organic pollutants. It acts as a funding entity whose implementing agencies comprise UN agencies and the World Bank. It is run with a small bureaucracy. It has achieved close relationships with the major MEAs. It permits some participation by NGOs, including at Council meetings. It has begun to gain more private sector involvement. The GEF operates transparently. With 176 member countries, the GEF has adopted a creative solution for solving the problem of internal democratic governance. The governing Council is reasonably-sized (32 members) with more from developing than developed countries. Even more innovatively, the members on the Council are appointed by a constituency of states for whom they represent (with some large states representing only themselves). As the GEF continues to mature, and its projects are evaluated, there may come a time when the perennial calls to 'strengthen UNEP' are replaced by a more apt proposal to *broaden* GEF.

3   The Paradigm of WEO as a Competitor

In the literature advocating a WEO, the rationale for the organization is described as promoting coherence within the regime, achieving economies and efficiencies, or carrying out cooperation with other international organizations (for example, Bergsten 1994, 364). Thus, much of the emphasis has been about what functions should be included within the WEO and what functions should be excluded.

Instead of designing a WEO with an eye only to internal coherence and external cooperation, the models for a WEO should better reflect an underlying goal of making the WEO a more effective *competitor* against other regimes. The notion that a WEO would be 'an institutional counterweight' to the trading system was a key insight in Esty's early analysis (see Esty 1994, 230; see also Esty 1999, 1560-61). Yet Esty, even while emphasizing the value of competition in other contexts, has not highlighted competition as a paradigmatic feature of the WEO. In my view, a WEO ought to champion environmental causes as it interacts with other international organizations such as the WTO, the World Bank, and the UN Development Programme.

The idea that the architecture of governance requires competition is an old one, going back to James Madison, if not earlier. In *Federalist Papers* No. 51, Madison explains that the United States Constitution should contrive the structure of the government so that its several constituent parts may, by their mutual relations, be the means of keeping each other in their proper places (Madison 1788). He further explains that in all subordinate distributions of power, the aim is to divide the several offices in such a manner as that each may be a check on the other. Although Madison writes about a national constitution, the same principle could apply to an international system or constitution.

Overlapping competence of agencies is a characteristic feature of international governance. Most analytical attention seems to go to managing the overlaps at different vertical levels of authority. Yet the horizontal overlaps are equally challenging, and require active efforts to seek coherence (see Sampson 2003). My point here is that while cooperation is one avenue to obtain coherence, it is not an exclusive one. Coherence can also be achieved through competition.

The need for competition is most apparent in the relationship between 'international environmental' and 'international economic' governance. In recent years, the WTO has climbed to a dominant position from which it seeks to insinuate its norms into other organizations. The claim is often made that WTO law trumps other bodies of law, and that environmental treaties need to conform to trade law. Since the WTO went into force in 1995, environmental treaty negotiations have been monitored carefully to make sure that they do not contravene WTO rules. This adversarial stance by the trading system has led to a 'chill' in environmental policy-making. Close observers of the WTO recognize that it suffers a 'superiority complex' (Pauwelyn 2003, 1177). Currently, the WTO is negotiating several environmental issues in the Doha Round. Although UNEP and several secretariats of multilateral environmental agreements have been invited to some of these negotiating sessions as ad hoc observers, the environment regime is powerless within the WTO to exercise any influence.

Because the trading regime is likely to give much more weight to commercial rather than ecological values, what has been missing is an evenly matched environment regime that can promote its norms in other arenas, and stand up to resist any overreaching by the WTO, or by new free trade agreements. Just as the WTO is now delving into trade-related environment policy, it would be appropriate for a WEO to delve into environment-related trade policy. For example, if a WEO existed now, it could be monitoring WTO negotiations on services to make sure that any new disciplines do not undercut environmental regulation. The recent proposal for a WEO included in the Heinrich Böll Foundation's *Jo'burg Memo* takes account of the value of horizontal competition at the international level. The *Memo* states that 'No system of checks and balances can be installed unless organizations like the ILO, the WHO, and the WTO are joined by an environmental organization of equal standing' (Sachs et al. 2002, 65).

It is beyond the scope of this article to present an organizational blueprint for a WEO that would prescribe a method of decision-making and means of enforcement. Good lessons can be learned from the GEF, and from the multilateral environmental agreements, which have pioneered institutional innovations (see Churchill and Ulstein 2000). The new network of environmental regulators is another important development that should be considered in designing a WEO.[16] In an era where hierarchy is giving way to networks, insights can come from any direction.

In calling for a WEO that could serve as a counterweight to the WTO and other institutions of economic governance, this article is not endorsing the WTO constitution as a template for a WEO. Certainly, the WTO has strengths that might be copyable. But the WTO also has many weaknesses, most notably its adherence to consensus-based decision-making that has recently arrested progress in the current Doha round of negotiations.

In its competition with the WTO, a WEO would have one important advantage. In contrast to the WTO, in which the vertical relationship with national trade ministries is one of supervision rather than cooperation, a WEO could develop a more cooperative association with national environment ministries, which themselves would be in a cooperative relationship with each other. Such environmental interdigitation toward common goals would be a feature not present in the trade regime, which is hard-wired for economic nationalism. Because the environment regime is so weak at the international level, most of the possibilities for fruitful vertical cooperation remain to be harvested. For example, improvements to

---

[16]   International Network for Environmental Compliance and Enforcement, available at http://www.incec.org.

environmental legal norms could occur through more systematic vertical borrowing (Wiener 2001). New ideas at the local level could be evaluated and, if successful, offered to other countries.

4   Conclusion

In penning this reflection, I am mindful that progress in ecological protection continues to occur, and that the tiny steps at Johannesburg may yield dividends not yet apparent. What worries me though is that the remarkable resilience of the biosphere is being taken for granted. Many opportunities to prevent a loss of natural resources are being missed.

I am also mindful that diversity within the environment regime can be valuable (Sand 2001, 297), and that many environmental tasks are disjoinable from others. Nevertheless, the fragmented nature of today's environmental governance defies organizational logic and perpetuates weak responses. If, over the next decade, UNEP is cabined to its present status and no better methods ensue for carrying out international environmental policy, then governments may fail to make much progress in responding to global challenges.

Supporters of a WEO should renew efforts to make the case for why organizational change can improve policies. In this article, I have addressed each aspect of the W-E-O, and pointed out the danger of allowing environmental governance to muddle along while economic governance grows stronger.

References

Bergsten, C.F. (1994), 'Managing the World Economy of the Future', in P.B. Kenen (ed.), *Managing the World Economy: Fifty Years after Bretton Woods*, Institute for International Economics, Washington, 341–74.

Bermann, G.A. (1994), 'Taking Subsidiarity Seriously: Federalism in the European Community and the United States', *Columbia Law Review*, vol. 94, 332–456.

Biermann, F. (2000), 'The Case for a World Environment Organization', *Environment,* vol. 42 (9), 23–31.

Biermann, F. (2001), 'The Emerging Debate on the Need for a World Environment Organization', *Global Environmental Politics*, vol. 1 (1), 45–55.

Biermann, F. (2005), 'The Rationale for a World Environment Organization', in F. Biermann and S. Bauer (eds), *A World Environment Organization. Solution or Threat for Effective International Environmental Governance?*, Ashgate, Aldershot, 117–144.

Carozza, P.G. (2003), 'Subsidiarity as a Structural Principle of International Human Rights Law', *American Journal of International Law*, vol. 97, 38–79.

Chambers, W.B. and J.F. Green (eds) (2004), *Reforming International Environmental Governance: From Institutional Limits to Innovative Solutions*, UNU Press, Tokyo.

Chayes, A. (1972), 'International Institutions for the Environment', in J. Lawrence Hargrove (ed.), *Law, Institutions and the Global Environment*, Oceana Publications, Dobbs Ferry, 1–26.

Churchill, R.R. and G. Ulstein (2000), 'Autonomous Institutional Arrangements in Multilateral Environmental Agreements: A Little-Noticed Phenomenon in International Law', *American Journal of International Law*, vol. 94, 623–59.

Dowdeswell, E. and S. Charnovitz (1997), 'Globalization, Trade and Interdependence', in M.R. Chertow and D.C. Esty (eds), *Thinking Ecologically. The Next Generation of Environmental Policy*, Yale University Press, New Haven.

Dua, A. and D.C. Esty (1997), *Sustaining the Asia Pacific Miracle*, Institute for International Economics, Washington.

Elliott, L. (2005), 'The United Nations' Record on Environmental Governance: An Assessment', in F. Biermann and S. Bauer (eds), *A World Environment Organization. Solution or Threat for Effective International Environmental Governance?*, Ashgate, Aldershot, 27–56.

Esty, D.C. (1994), *Greening the GATT*. Institute for International Economics, Washington.

Esty, D.C. (1999), 'Toward Optimal Environmental Governance', *New York University Law Review*, vol. 74, 1495–574.

Esty, D.C. (2001), 'A Term's Limits', *Foreign Policy*, September/October, 74–75.

Esty, D.C. and M.H. Ivanova (2002), 'Revitalizing Global Environmental Governance: A Function-Driven Approach', in Esty and Ivanova (eds), *Global Environmental Governance. Options and Opportunities*, Yale School of Forestry and Environmental Studies, New Haven (Conn.), 181–203.

French, H. (2003), 'New Leadership from the South', *WorldWatch*, July–August, 2.

G8 Environment Ministers Communiqué (2003), available at http://www.g7.utoronto.ca/-environment/2003paris/env_communique_april_2003_eng.html.

Gaines, S.E. (2003), 'The Problem of Enforcing Environmental Norms in the WTO and What To Do About It', *Hastings International and Comparative Law Review*, Spring 2003, 321–85.

Gardner, R. N. (1974), 'The Hard Road to World Order', *Foreign Affairs*, April 1974, 556–76.

German Advisory Council on Global Change (2001), *New Structures for Global Environmental Policy*, Earthscan, London.

Global Accountability Report (2003), *Power Without Accountability?*, One World Trust, [available at http://www.oneworldtrust.org/].

Gray, K.R. (2003), 'World Summit on Sustainable Development: Accomplishments and New Directions?' *International and Comparative Law Quarterly*, vol. 52, 256–68.

Gutman, P. (2003), 'What Did WSSD Accomplish? An NGO Perspective', *Environment*, vol. 45 (2), 21–28.

Haas, P.M. (2004), 'Addressing the Global Governance Deficit', unpublished paper.

Hierlmeier, J. (2002), 'UNEP: Retrospect and Prospect: Options for Reforming the Global Environmental Governance Regime', *Georgetown International Environmental Law Review*, vol. 14, 767–805.

High-Level Panel on Financing for Development, Report, in UN General Assembly, A/55/1000 (26 June 2001).

Holdgate, M. (1999), *The Green Web*, Earthscan, London.

Holliday Jr., C.O., S. Schmidheiny and P. Watts (2002), *Walking the Talk. The Business Case for Sustainable Development,* Greenleaf Publishing, Sheffield.

Johannesburg Declaration and Plan of Implementation (2002), available at http://-www.johannesburgsummit.org/.

Juma, C. (2000), 'The Perils of Centralizing Global Environmental Governance', *Environment Matters* (World Bank), 2002, 13–15, [available at http://www-wds.worldbank.-org/servlet/WDSContentServer/WDSP/IB/2000/11/04/000094946_0010130548127/Rendered/PDF/multiopage.pdf].

Keohane, R.O., P.M. Haas and M.A. Levy (1993), 'The Effectiveness of International Environmental Institutions', in P.M. Haas, R.O. Keohane and M.A. Levy (eds), *Institutions for the Earth*, MIT Press, Cambridge (Mass.), 3–24.

Kirton, J. (2005), 'Generating Effective Global Environmental Governance: The North's Need for a WEO', in F. Biermann and S. Bauer (eds), *A World Environment Organization. Solution or Threat for Effective International Environmental Governance?*, Ashgate, Aldershot, 145–172.

Levy, M.A., R.O. Keohane and P.M. Haas, (1993), 'Improving the Effectiveness of International Environmental Institutions', in Haas, Keohane and Levy (eds), *Institutions for the Earth*, MIT Press, Cambridge (Mass.), 397–426.

Lodefalk, M. and J. Whalley, (2002), 'Reviewing Proposals for a World Environmental Organisation', *The World Economy*, vol. 25, 601–17.

MacDonald, G. J. (2003), 'Environment: Evolution of a Concept', *Journal of Environment and Development*, vol. 12, 151–76.

Madison, J. (1788), 'No. 51', available at http://memory.loc.gov/const/fed/fedpapers.html.

Malmö Declaration (2000), available at www.unep.org/malmo/malmo_ministerial.htm.

Marshall, D. (2002), 'An Organization for the World Environment: Three Models and Analysis', *Georgetown International Environmental Law Review*, vol. 15, 79–03.

McNeill, J.R. (2000), *Something New Under the Sun. An Environmental History of the Twentieth-Century World*, W. W. Norton and Company, New York.

Najam, A. (2002), 'Why We Don't Need a New International Environmental Organization' (Draft), [available at http://www.bu.edu/cees/research/workingp/pdfs/Najam-GEO-SDPI.pdf].

Najam, A. (2005), 'Neither Necessary, Nor Sufficient: Why Organizational Tinkering Won't Improve Environmental Governance', in F. Biermann and S. Bauer (eds), *A World Environment Organization. Solution or Threat for Effective International Environmental Governance?*, Ashgate, Aldershot, 235–256.

Nanda, H. S. (2003), 'India Convinces UNEP to Eliminate Asian "Brown Cloud" from Meeting Agenda', *International Environment Reporter*, vol. 26, 246.

Newell, P. (2002), 'A World Environmental Organisation: The Wrong Solution to the Wrong Problem', *The World Economy*, vol. 25, 659–71.

Oberthür, S. and T. Gehring (2005), 'Reforming International Environmental Governance: An Institutional Perspective on Proposals to Establish a World Environment Organization', in F. Biermann and S. Bauer (eds), *A World Environment Organization. Solution or Threat for Effective International Environmental Governance?*, Ashgate, Aldershot, 205–234.

Ostry, S. (2001), 'The WTO and International Governance', in K.G. Deutsch and B. Speyer (eds), *The World Trade Organization Millennium Round*, Routledge, London, 285–94.

Palmer, G. (1992), 'New Ways to Make International Environmental Law', *American Journal of International Law*, vol. 86, 259–83.

Pauwelyn, J. (2003), 'What to Make of the WTO Waiver for "Conflict Diamonds": WTO Compassion of Superiority Complex?' *Michigan Journal of International Law*, vol. 24, 1177–207.

Pius XI, Pope (1931), 'Quadragesimo Anno, on Reconstruction of the Social Order', available at http://www.vatican.va/holy_father/pius_xi/encyclicals/documents/hf_p-xi_enc_-19310515_quadragesimo-anno_en.html

Raustiala, K. (2002), 'The Architecture of International Cooperation: Transgovernmental Networks and the Future of International Law', *Virginia Journal of International Law,* vol. 43, 1–92.

Rio Declaration on Environment and Development (1992), in United Nations, *Earth Summit Agenda 21,* United Nations, New York.

Runge, C.F. (1994), *Freer Trade, Protected Environment,* Council on Foreign Relations, New York.

Runge, C.F. (2001), 'A Global Environment Organization (GEO) and the World Trading System', *Journal of World Trade,* vol. 35, 399–426.

Sachs, W. et al. (2002), 'The Jo'burg Memo. Fairness in a Fragile World', Heinrich Böll Foundation, available at http://www.joburgmemo.org/.

Sampson, G.P. (2003), *Is there a Need for Restructuring the Collaboration among the WTO and UN Agencies so as to Harness Their Complementarities',* Paper presented at the European University Institute, June 2003.

Sand, P.H. (2001), 'Environment: Nature Cooperation', in P.J. Simmons and C. de J. Oudraat (eds), *Managing Global Issues. Lessons Learned,* Carnegie Endowment for International Peace, Washington, 281–309.

Shadow G–8 (2003), 'Restoring G–8 Leadership of the World Economy', May 1993, available at http://www.ifri.org/files/policy_briefs/WP_SHADOW_G8.pdf.

Simonis, U.E. and T. Brühl (2002), 'World Ecology—Structures and Trends', in P. Kennedy, D. Messner and F. Nuscheler (eds), *Global Trends and Global Governance,* Pluto Press, London, 97–124.

Speer, L.J. (2003), 'G-8 Environment Ministers Pledge Better Maritime Safety, Disagree on Timing', *BNA Daily Environment Report,* 29 April 2003, A-4.

Speth, J.G. (2002a), 'The Global Environmental Agenda: Origins and Prospects', in Esty and Ivanova (eds), *Global Environmental Governance. Options and Opportunities,* Yale School of Forestry and Environmental Studies, New Haven, 11–30.

Speth, J.G. (2002b), 'Attacking the Root Causes of Global Environmental Deterioration', *Environment,* vol. 44 (7), 16–25.

Speth, J.G. (2004), *Red Sky at Morning,* Yale University Press, New Haven.

Stein, R. (2003), 'WHO Gets Wider Power to Fight Global Health Threats', *Washington Post,* 28 May 2003.

Stone, C.D. (1993), *The Gnat is Older than Man,* Princeton University Press, Princeton.

Streck, C. (2001), 'The Global Environment Facility–A Role Model for International Governance?', *Global Environmental Politics,* vol. 1 (2), 71–94.

Supachai, P. (2001), 'The Evolving Multilateral Trading System in the New Millennium', *George Washington International Law Review,* vol. 33, 419–49.

Tarlock, A.D. (2001), 'Ideas without Institutions: The Paradox of Sustainable Development', *Indiana Journal of Global Legal Studies,* vol. 9, 35–49.

Tolba, M.K. (1998), *Global Environmental Diplomacy,* MIT Press, Cambridge (Mass.).

Turner, T. (2000), 'Critical Crossroads', *Our Planet,* vol. 10 (5), 4–5.

UNDP [United Nations Development Programme], UNEP [United Nations Environment Programme], World Bank, World Resources Institute (2003), *World Resources 2002–2004,* World Resources Institute, Washington.

UNEP [United Nations Environment Programme] (2002), *UNEP in 2002,* available at http://www.unep.org/Evaluation/default.htm.

Vergés Bausili, A. (2002), 'Rethinking the Methods of Dividing and Exercising Powers in the EU: Reforming Subsidiarity and National Parliaments' Jean Monnet Working Paper 9/02, available at www.jeanmonnetprogram.org/papers.

von Moltke, K. (2001), 'The Organization of the Impossible', *Global Environmental Politics,* vol. 1 (1), 23–28.

von Moltke, K. (2005), 'Clustering International Environmental Agreements as an Alternative to a World Environment Organization', in F. Biermann and S. Bauer (eds), *A World Environment Organization. Solution or Threat for Effective International Environmental Governance?,* Ashgate, Aldershot, 175–204.

Whalley, J. and B. Zissimos (2001), 'What Could a World Environmental Organization Do?' *Global Environmental Politics,* vol. 1 (1), 29–34.

Whalley, J. and B. Zissimos (2002), 'Making Environmental Deals: The Economic Case for a World Environmental Organization', in D. Esty and M. Ivanova (eds), *Global Environmental Governance. Options and Opportunities,* Yale School of Forestry and Environmental Studies, New Haven (Conn.), 163–80.

Wiener, J.B. (2001), 'Something Borrowed for Something Blue: Legal Transplants and the Evolution of Global Environmental Law', *Ecology Law Quarterly,* vol. 27, 1295–371.

Wolfrum, R. and N. Matz (2003), *Conflicts in International Environmental Law,* Springer, Berlin.

World Commission on Environment and Development (1987), *Our Common Future,* Oxford University Press, Oxford.

Worldwatch (2003), *State of the World 2003,* W. W. Norton and Company, New York.

Chapter 5

The Rationale for a World Environment Organization

Frank Biermann*

1   Introduction

Most observers agree on the need to reform and strengthen the current system of global environmental governance. This chapter argues that one element of such reform would be to upgrade the United Nations Environment Programme (UNEP) to a specialized agency of the United Nations. This proposal is motivated by the imbalance between UNEP, currently a modest UN sub-programme, versus the plethora of influential intergovernmental organizations in the fields of labour, shipping, agriculture, communication, culture, or economic policy. As a mere programme, for example, UNEP has no right to adopt treaties or any regulations upon its own initiative, it cannot avail itself of any regular and predictable funding, and it is subordinated to the UN Economic and Social Council. UNEP's staff hardly exceeds 450 professionals—a trifle compared to its national counterparts such as the German Federal Environment Agency with over 1,000 employees and the United States Environmental Protection Agency with a staff of roughly 19,000. This situation has led to a

*   I thank Steffen Bauer, Aarti Gupta and four anonymous reviewers for valuable comments. Funding for this research by Volkswagen Foundation, Hanover, is gratefully acknowledged.

variety of proposals to grant the environment what other policy areas have long had: a strong international agency with a sizeable mandate, significant resources and sufficient autonomy.

I call this new agency—in line with the World Health Organization (WHO) or the World Meteorological Organization (WMO)—a 'World Environment Organization' (WEO). The debate on a WEO has been underway for some time, dating back more than thirty years to George Kennan's (1970) proposal for an international environmental agency (see the introductory chapter by Bauer and Biermann in this volume for an overview). I have taken part in the debate for eight years; this chapter presents the basic three objectives that I believe a world environment organization could achieve, and discusses five major issues that currently stand at the centre of the debate. The article draws on, but also differs from my previous writing in that it reflects some of the criticisms that have been raised about a WEO, including by contributors to this volume.[1]

In a nutshell, I propose to maintain the current system of issue-specific international environmental regimes while strengthening environmental protection by *upgrading UNEP* from a mere UN programme to a full-fledged international organization with increased financial and staff resources and enhanced competences and legal mandate. In this model, a world environment organization would function among the other international institutions and organizations, whose member states might then be inclined to shift some competencies related to the environment to the new world environment organization. In particular, the new organization would provide a venue for the co-location and eventually joint administration of the myriad convention secretariats. The organization would also have its own budget, based on assessed contributions by member states, and it could make use of future innovative financial mechanisms, for example revenues from emissions trading regimes. Additional financial and staff resources could be devoted to the fields of awareness raising, technology transfer and the provision of environmental expertise to international, national and sub-national levels. The elevation of

---

[1]   Parts of this chapter draw on Biermann 2000, 2001a, and 2002b.

UNEP to a world environment organization of this type could be modelled on the WHO or the International Labour Organization (ILO), that is, independent international organizations with their own membership.

If governments agreed on establishing a world environment organization as a specialized UN agency, this body would be based on a constitutive legal instrument that would require ratification of a certain number of states to become effective. Hence the creation of such an organization would not require the legal acquiescence of all nations, and it would have autonomy over its own organizational design. A decision by governments in the UN General Assembly would be needed to formally abolish UNEP and to transfer its staff and assets to the new agency.

Upgrading UNEP to a specialized UN agency would follow the long-standing policy of functional specialization within the UN system, with the 'United Nations Organization' as the focal point among numerous independent organizations for specific issues, such as food and agriculture (FAO, established in 1945); education, science, and culture (UNESCO, 1945); health (WHO, 1946); civil aviation (ICAO, 1944); or meteorology (WMO, 1947). While some specialized organizations are much older than the United Nations itself (for instance the Universal Postal Union, created in 1874), most were founded simultaneously with the establishment of the United Nations, since it was felt at that time that the vast number of issues in the economic, social or technical fields would 'over-stretch' the world body. Environmental problems, however, were no concern in 1945, with the term 'environment' not even appearing in the UN Charter. It was only in 1972 that UNEP was set up as a mere programme, without legal personality, without budget, and—according to its founding instruments—with only a 'small secretariat', which is no comparison to the other specialized organizations that can avail themselves of more resources and hence influence.

Given the multitude of reform pitches, I believe that my proposal is a moderate proposition. It even seems to be agreeable to some of the outspoken opponents of a world environment organization. Adil Najam, for example, proposes in his contribution to this volume (p. 248) 'to convert UNEP into a specialized agency (as opposed to a "Programme") with the concomitant ability to raise and decide its own budget'. This is more or less what I propose here as a world environment organization.

I hence differ from some of the more radical approaches in the literature that demand, for example, the abolishment of major existing agencies such as the World Meteorological Organization, the creation of a new agency with enforcement power—e.g. through trade sanctions—or the creation of a new agency *in addition to* UNEP, which would remain but transfer many of its functions to the new organization (Esty 1994, 1996; Haas, Kanie and Murphy, forthcoming). Most of these radical designs appear both unrealistic and undesirable: abolishing major UN agencies has been rare in post-1945 history and seems politically unfeasible or unnecessary today. Strong enforcement mechanisms, such as trade sanctions, tend to unfairly focus on less powerful developing countries while leaving larger industrialized countries unaffected (Biermann 2001b). Establishing a new agency *in addition* to UNEP—as also proposed by Kirton in this volume—might even create new coordination problems while attempting to solve others.

In the following, I lay out three core functions that a world environment organization should fulfil. Section 3 then addresses five major issues of current discussion.

2   Three Core Functions of a World Environment Organization

Better Coordination of Global Environmental Governance

Upgrading UNEP to a UN specialized agency—and thus to a WEO—could ameliorate, first, the coordination deficit in the international governance architecture that results in substantial costs and sub-optimal policy outcomes. Since 1972, when UNEP was set-up, the increase in international environmental regimes has led to a considerable fragmentation of the entire system. Norms and standards in each area of environmental governance are created by distinct legislative bodies—the conferences of the parties to various conventions—with little respect for repercussions and links with other fields. While the decentralized negotiation of rules and standards in separate functional bodies may be defensible, this is less so regarding the organizational fragmentation of the various convention secretariats, which have evolved into quite independent bureaucracies with strong centrifugal tendencies (Biermann and Bauer 2004a and forthcoming). In addition, most specialized organizations and bodies, such as the UN Food and Agriculture Organization (FAO) or the UN Organization for Industrial Development (UNIDO), have initiated their own environmental programmes independently from each other and with little policy coordination among themselves and with UNEP. The prevailing situation at the international level might come close, if compared to the national level, to the abolishment of national environment ministries and the transfer of their programmes and policies to the ministries of agriculture, industry, energy, economics or trade: a policy proposal that would not find many supporters in most countries. For good reasons, there are no functionally different secretariats for the many conventions on labour or on trade, which are administered instead by single specialized organizations, the International Labour Organization (ILO) and the World Trade Organization (WTO), respectively.

This problem is well known. The attempt to network individual organizations, programmes and offices has been ongoing since 1972, when a first coordinating body was set up within the UN. This and its successors, however, have lacked the legal authority to overcome the special interests of individual departments, programmes and convention secretariats. For global environmental policy, no central anchoring point exists that could compare to WHO or ILO in their respective fields. Instead, there is an overlap in the functional areas of several institutions. As reaffirmed by environment ministers represented in the UNEP expert group on international environmental governance, there is a 'compelling rationale for a comprehensive effort at rationalizing, streamlining and consolidating the present system [of] multilateral environmental agreements' (UNEP 2001a, para. 9). An international centre with a clear strategy to ensure worldwide environmental protection is thus the need of the hour. Just as within nation states, where environmental policy was strengthened through introduction of independent environmental ministries, global environmental policy could be made stronger through an independent world environment organization that helps to contain the special interests of individual programmes and organizations and to limit duplication, overlap and inconsistencies.

In addition to the co-location of secretariats of multilateral environmental treaties under a WEO umbrella, governments could chose to empower the new agency to also coordinate existing multilateral environmental regimes and new environmental agreements, generally by a decision of the respective conferences of the parties. The constitutive treaty of the organization could provide general principles for multilateral environmental treaties as well as coordinating rules that govern the organization and its relationship with the issue-specific environmental regimes.

Following WTO usage, environmental regimes covered by the WEO could be divided into 'multilateral' and 'plurilateral' environmental agreements. For 'multilateral' agreements, ratification would be compulsory for any new member of the organization, while 'plurilateral' agreements would still leave members the option to remain outside. The multilateral agreements would thus form the 'global environmental law code' under the world environment organization, with the existing conferences of the parties—say, to the Montreal Protocol on Substances that Deplete the Ozone Layer—being transformed *de jure* or *de facto* into sub-committees under the WEO Assembly. This would enable the WEO Assembly to develop a common reporting system for all multilateral environmental agreements (e.g., an Annual National Report to the WEO); a common dispute settlement system; mutually agreed guidelines that could be used—based on an inter-agency agreement—for the environmental activities of the World Bank and for environmentally-related conflicts regulated under the WTO dispute settlement system; as well as a joint system of capacity-building for developing countries along with financial and technological transfer.

Streamlining of environmental secretariats and negotiations into one body would especially increase the voice of the South in global environmental negotiations. The current system of organizational fragmentation and inadequate coordination causes special problems for developing countries. Individual environmental agreements are negotiated in a variety of places, ranging—for example in ozone policy—from Vienna to Montreal, Helsinki, London, Nairobi, Copenhagen, Bangkok, Nairobi, Vienna, San José, Montreal, Cairo, Beijing and Ouagadougou. Recent conferences on climate change, as another example, were hosted in a circular movement covering four continents (Berlin, Geneva, Kyoto, Bonn, Buenos Aires, The Hague, Marrakech, New Delhi, Milan and again Buenos Aires). This nomadic nature also characterizes most sub-committees of environmental conventions.

Developing countries lack the resources to attend all these meetings with a sufficient number of well-qualified diplomats and experts. Often, even larger developing countries—and occasionally developed countries too[2]—need to rely on their local embassy staff to negotiate highly complex technical regulations on the environment. The task of adjusting and amending the highly technical lists of chemicals controlled under the Montreal Protocol in Helsinki 1989, for example, was entrusted to the Indian ambassador to Finland (Rajan 1997). This system of a 'travelling diplomatic circus' distinguishes environmental governance from other policy fields, where negotiations are usually held within the assembly of an international agency at its seat. The creation of a world environment organization could thus help developing countries to build up specialized 'environmental embassies' at the seat of the new organization. This would reduce their costs and increase their negotiation skills and hence influence.[3] The same could be said for non-governmental organizations, which could participate in global negotiations within the WEO Assembly and its committees at lower costs.

It has been argued that global public policy networks would be the answer to the coordination problem in global environmental governance, not a new UN agency (e.g. Najam 2003 and in this volume). This contrast of networks versus agency, however, is flawed and stems from an incomplete

---

[2]   For example, industrialized countries such as The Netherlands, Germany or Sweden do not maintain permanent representations at the headquarters of the International Maritime Organization, along with many developing countries.

[3]   UNEP (2001a) lists, for example, the sizeable benefits of simply hosting conferences of the parties back to back in the same location. Much larger benefits could be reaped if *all* meetings were, as a rule, held in one place. See UNEP 2001a, para. 88: 'The most obvious way is to host the conference of the parties of conventions within a cluster back to back in the same location. The most obvious costs are for conference facilities, which are usually covered by the host Government. Additionally there are considerable costs borne by the participating delegates, observers and the media to cover airfares and accommodation. Finally, there are costs related to setting up temporary offices and communication infrastructure. Additionally there will almost certainly be cost savings for the secretariats by opting to pool resources when hosting two conferences of the parties back to back. Similarly, Governments would be able to realize cost-efficiency gains by maintaining the same communications infrastructure for both conventions. In addition, carefully planned back-to-back events would facilitate greater substantial cross-cutting negotiations, and would probably weed out substantive inconsistencies or grey area issues that still exist within the international regime of international law'.

reading of the WEO literature. Most global public policy networks include international organizations, and an upgraded UNEP would continue to be part of these networks, yet be stronger and more active through a strengthened mandate and improved resources. Instead of superseding public policy networks, a WEO would be complementary to the existing networks that have evolved around specific issues; it could even spearhead and lead new networks on new issues as they come onto the agenda. The World Commission on Dams, as an example for a prominent global public policy network, has been initiated by the World Bank (Dingwerth, forthcoming)—this initiative could as well have been the task of a WEO.

Better Development and Implementation of International Environmental Law

If UNEP were upgraded to a UN specialized agency—a WEO—the body would also be better poised to support regime-building processes, especially by initiating and preparing new treaties. The ILO can serve as a model here. ILO has developed a comprehensive body of 'ILO conventions' that come close to a global labour code. In comparison, current global environmental policy is far more disparate and cumbersome in its norm-setting processes. It is also riddled with various disputes among the UN specialized organizations regarding their competencies, with UNEP in its current form being unable to adequately protect environmental concerns.

A specialized UN organization could also approve—by qualified majority vote—certain regulations, which are then binding on all members, comparable to article 21 and 22 of the WHO Constitution.[4] Likewise, the

---

4   Cf. WHO Constitution: '*Article 21:* The Health Assembly shall have authority to adopt regulations concerning: (*a*) sanitary and quarantine requirements and other procedures designed to prevent the international spread of disease; (*b*) nomenclatures with respect to diseases, causes of death and public health practices; (*c*) standards with respect to diagnostic procedures for international use; (*d*) standards with respect to the safety, purity and potency of biological, pharmaceutical and similar products moving in international commerce; (*e*) advertising and labelling of biological, pharmaceutical and similar products moving in international commerce. *Article 22:* Regulations adopted pursuant to Article 21 shall come into force for all Members after due notice has been given of their adoption by the Health Assembly except for such Members as may notify the Director-General of rejection or reservations within the period stated in the notice.'

WEO Assembly could adopt draft treaties which have been negotiated by sub-committees under its auspices and which would then be opened for signature within WEO headquarters. The ILO Constitution, for example, requires its parties in article 19:5 to submit, within one year, all treaties adopted by the ILO General Conference to the respective national authorities (such as the parliament) and to report back to the organization on progress in the ratification process. This goes much beyond the powers of the UNEP Governing Council, which can *initiate* intergovernmental negotiations, but cannot *adopt* legal instruments on its own.

Apart from regime-building and norm-setting, a WEO could also improve the overall implementation of international environmental policy, for example by a common comprehensive reporting system on the state of the environment and on the state of implementation in different countries, as well as by stronger efforts in raising public awareness. At present, several environmental regimes require their parties to report on their policies, and a few specialized organizations collect and disseminate valuable knowledge and promote further research.[5] Yet there remains a sizeable lack of coordination, bundling, processing and channelling of this knowledge in a policy-oriented manner. Most conventions still have different reporting needs and formats, with a certain amount of duplication. The current system is burdensome especially for developing countries, since the myriad reporting systems siphon off administrative resources that governments could use for other purposes.

All reporting requirements could, however, easily be streamlined into one single report to be dispatched to one single body, such as a WEO, which would be based on a legal agreement specifying the reporting processes and needs. Instead of adding another layer of bureaucracy, as often surmised by reform opponents, a WEO would thus provide a level of

---

5   Peter Haas (2002), for example, has sampled 48 multilateral environmental agreements that call for environmental quality monitoring; in 81 per cent of these cases, the submission of a report is mandatory. 17 per cent of these multilateral environmental agreements require a new report each year, which places additional burden on understaffed ministries in many developing countries.

streamlining and harmonization that would reduce the current administrative burden, in particular for developing countries.

Improved Financial and Technology Transfer to the South

Upgrading UNEP to a UN specialized agency—a WEO—could also assist in the build-up of environmental capacities in developing countries. Strengthening the capacity of developing countries to deal with global and domestic environmental problems is one of the essential functions of international environmental regimes (e.g., Keohane and Levy 1996; Biermann 1997). Yet the current organizational setting for financial North-South transfers suffers from an ad-hocism and fragmentation that does not fully meet the requirements of transparency, efficiency and participation of the parties involved. At present, most industrialized countries strive for a strengthening of the World Bank and the affiliate Global Environment Facility (GEF), to which they will likely wish to assign most financial transfers. Yet many developing countries continue to perceive the World Bank as a Northern-dominated institution ruled by decision-making procedures based on contributions. Though the GEF has been substantially reformed in 1994 and is governed together with UNEP and UNDP, it still faces criticisms from the South.

An alternative could be to move the tasks of overseeing capacity building and financial and technological assistance for global environmental policies to an independent body that is specially designed to account for the distinct character of North-South relations in global environmental policy, that could link the normative and technical aspects of financial and technological assistance, and that is strong enough to overcome the fragmentation of the current multitude of single funds. Such a body could be a WEO. The WEO could be empowered to coordinate various financial mechanisms and administer the funds of sectoral regimes in trust. In addition, a WEO could host the clean development mechanism and the clearinghouse for the future emissions trading scheme under the Kyoto Protocol to the UN Framework Convention on Climate Change, which would reduce bureaucratic overlap, increase efficiency and assist in preventing conflicts with other, non-climate related environmental problems. The Kyoto mechanisms could also contribute by user fees to the financing of the new organization.

Finally, the new body could be used for any future scheme of automatic financial mechanisms, such as global user fees on air and maritime transport, as has been recently proposed, for example, by the German Environment Minister.[6] Last but not least, an organization, as opposed to a programme, could allow for a system of regular, predictable and assessed contributions of members, instead of voluntary contributions generated at unpredictable 'pledging conferences', as is currently the case with UNEP.

There is no need to set up new large financial bureaucracies. Instead, a WEO should still make use of the expertise of World Bank or the UN Development Programme (UNDP), including their national representatives in developing countries. By designating a WEO as a central authoritative body for the various financial mechanisms and funds, however, the control of developing countries over implementation could be strengthened, without giving away the advantages of technical expertise and knowledge of existing organizations. The norm-setting functions of the GEF, for example regarding the criteria for financial disbursement, could be transferred to the WEO Assembly, in a system that would leave GEF the role of a 'finance ministry' under the overall supervision and normative guidance of the WEO Assembly. This would unite the economic and administrative expertise of the GEF secretariat with the 'legislative' role of a world environment organization.[7]

---

[6]   In a recent speech, the German Federal Minister for the Environment, Nature Conservation and Nuclear Energy, Jürgen Trittin, stated: '[T]he protection of global goods requires [...] additional funds. Protecting the climate, biodiversity and international water resources costs money, particularly in developing countries. For this reason, a global environmental institution should not be solely dependent on contributions from its member states. We should identify financial sources on the basis of the polluter-pays principle. Global goods are valuable, therefore they should not be used for free, as in air and maritime transport, for example. A price on these goods would reduce the level of use to a more tolerable one and would make global environmental institutions less dependent on contributions from nation states' (Trittin 2002, 12).

[7]   Cf. Andler (forthcoming) for a recent analysis of the GEF secretariat.

3   Core Issues of Debate

This section addresses some of the major issues and concerns in the debate on a world environment organization. Five issues are discussed: whether such an agency would focus only on global issues or also on local issues (in particular with regard to developing countries); what role civil society could play in global governance, including within a world environment organization; whether such a body should focus on environmental protection or rather on sustainable development; whether such an agency would be in the interest of developing countries; and whether the clustering of multilateral environmental agreements is an alternative to a WEO.

Global Commons versus World Problems

The creation of a new agency will require the delineation of its mandate, especially whether it will cover all environmental problems or just a sub-set, the so-called 'global' environmental problems. I believe that a WEO will only be functional and universally acceptable if it covers environmental concerns at all levels. Some other writers, however—most explicitly Daniel C. Esty and Maria Ivanova—have argued for a 'Global Environmental Organization' (GEO) that would exclusively deal with what they conceive of as 'global environmental problems' (Esty and Ivanova 2001; also Esty 1994, 1996). While they have not yet listed in detail the environmental issues that should be addressed by a GEO, they contend that local problems—such as local air pollution, soil degradation or water poisoning—must not be part of a GEO.[8]

---

8   Esty and Ivanova (2001) state that they 'draw a conscious distinction between "global" environmental concerns and the environmental problems of the "world", which include many issues that span the globe but have only local impact'. They do not define which environmental concerns they see as global and where they draw the line between local and global components. Deforestation, for example, is both local—when it comes to soil degradation—and global, when it comes to global climate change. River pollution is part of a global problem through the degradation of coastal waters and, to a much lesser extent, of the high seas. In a later section of their article, they write about 'inherently global issues, e.g., those affecting the atmosphere, biodiversity, and the oceans', without elaborating on the question of whether this includes issues such as local farming practices, which affect at

This GEO concept is technically problematic, potentially unfair, and difficult to implement. First, the terms 'global environmental problems' or 'global commons' are hard to define in a legal-political context. Forests, for example, have been mentioned as a global common owing to their environmental functions in the earth system, but most developing countries would object to notions of limited sovereignty in this field. If the atmosphere were politically or legally defined as a 'global common', what would be the result for national energy policies, or traffic policies, in Uzbekistan, Uganda or the United States—would they be seen as global or as local issues? Consequently, the adjective 'global' has not been used to denote an international agency, with the notable exception of the Global Environment Facility, which expressly excludes local problems and has thus been criticized by developing countries (Ehrmann 1997; Fairman 1996; Jordan 1994).

A second problem is that UNEP addresses at present all forms of environmental problems, from the local to the global. The creation of a GEO, based on UNEP, would thus either entail the restriction of the current universal mandate of UNEP, or it would require the establishment of some parallel international entity for local environmental issues. A number of successful UNEP programmes, such as the UNEP Regional Seas Programme, would entirely fall out of the purview of such a 'global' organization. It does not seem unlikely that this development would create a two-tier, if not 'two-class' international organizational structure: first, a strong 'Global' Environment Organization with worldwide reach, significant financial resources and the support of industrialized countries, which deals with issues of immediate concern for the North, such as climate change, loss of biodiversity or ozone depletion; and second, a weak, if not non-existent, international mechanism for the local environmental problems of developing countries, ranging from water pollution to indoor air pollution.[9]

---

the same time the climate (e.g., methane emissions), biodiversity and the seas (river pollution).

[9] Esty and Ivanova (2001), for example, explicitly state (at p. 13) that 'activities aimed at local issues (provision of drinking water, wastewater treatment, land management, air pollution control, etc.) should be undertaken by national governments supported by UNDP, the World Bank, and other development entities'. This would effectively leave

This seems hardly acceptable for developing countries, and it would, in the long run, do little for the environment. The prevalent Southern distrust in this debate is mirrored in a recent UNEP report on 'convention clustering' which placed the conventions on climate and ozone depletion—presumably prime candidates for a 'global common issue'—not in a cluster of atmospheric issues (which is absent), but in a cluster of 'sustainable development conventions', indicating the special status which developing countries bestow on the socio-economic implications of the climate issue (UNEP 2001a, para. 25).

In sum, the creation of a 'Global' Environment Organization, as opposed to a world organization, would be tantamount to the 'downsizing', if not the abolishment of most efforts and programmes of the international community to tackle local environmental pollution and degradation in Africa, Asia and Latin America. When in the preparatory process to the Johannesburg a group of environmental ministers was set up to find ways to strengthen the UN system in this area, developing countries objected to the use of the term 'global environmental governance', but opted instead for 'international environmental governance'. If a new international UN agency on environmental issues were to be created, it would thus seem pivotal to structure it as a 'World Environment Organization', analogous to the WHO or the World Food Programme or other agencies with a universal mandate that includes local problems.[10]

---

policies on the most pressing environmental problems in the South to their own national governments, whereas global issues—that also affect industrialized countries—would be elevated to a stronger and more visible international status. For developing countries, such differentiation is hardly acceptable.

[10] Of course, as argued by Charnovitz (2002), one could agree on defining an organization with a universal mandate that includes local and transnational issues, and still keep the acronym GEO, which has, admittedly, the advantage of positive public associations with the ancient Greek word for (the goddess) Earth. It could well be possible to negotiate a constitutive legal agreement for an organization that would include local issues, but would bear the title GEO because of the superior connotations of the term. The adjective 'global' in GEO would then simply take a different meaning from its use in the GEF. For the current debate, however, it seems advisable to maintain the differentiation between 'world' and 'global' environment organization, since fundamentally different concepts are implied.

Civil Society versus Intergovernmental Agencies

It is not uncommon in the literature to juxtapose the WEO debate with the need to strengthen the role of civil society actors. UNEP, the Commission on Sustainable Development or the conferences of the parties to the major environmental conventions have all been important venues for the influence of environmentalists, scientists or business representatives. There is no reason to believe that upgrading UNEP to a specialized agency would decrease this influence.

Instead, a new body would allow governments and civil society actors to institutionalize the influence of non-state actors in a way that would make it stronger, but also more balanced. This is important since developing countries generally object to increases in the influence of non-governmental organizations in international fora because they view these groups as being more favourable to Northern agendas, perspectives and interests. Indeed, most associations are headquartered in industrialized countries, and most funds donated to their cause stem from Northern organizations, both public and private. In quite a few cases, this situation influences the agenda of these groups, which are more accountable to Northern audiences than to elected Southern authorities (South Centre 1996). The same holds for international scientific networks, which are largely sustained by Northern scientists and Northern research funds. Here, too, it has been argued that 'international science', in particular in its social science parts, is structured in a way that favours Northern perspectives and is hence not taken at face value by Southern constituencies (Biermann 2002a with further references).

These suspected biases in the work of non-governmental actors, however, should not lead to a decrease in the participation of civil society, but rather to the setting-up of mechanisms that ensure a balance of opinions and perspectives. A world environment organization, with a strong and influential WEO Assembly, could further this goal. It would provide a highly visible locus for the activity of non-governmental groups while offering at the same time mechanisms in its decision-making process that would institutionalize the influence of non-governmental groups in a way that guarantees the balance of views and interests.

A general model for achieving this balance—though not necessarily fitting as a one-to-one blueprint—is the ILO decision-making procedure (see also Charnovitz, this volume). Here, each member state is represented with four votes, two of which are assigned to governments and one each to business associations and labour unions. An ILO-type procedure would address the basic problem of a 'global civil society', namely that environmental groups can often not adequately compete with the financial clout of business associations, and that non-governmental organizations of developing countries lack standing vis-à-vis the financially well-endowed non-governmental organizations of industrialized countries. An ILO-type structure would thus grant business interests and environmental interests at least formally equal rights, and it would guarantee that the Southern non-governmental associations would have clout in accordance with the population represented by them. Of course, a number of problems attach to the transferral of such a procedure to international environmental policy. There are as yet only few federations of environmental organizations that convincingly represent their entire national clientele, and many smaller developing countries lack non-governmental organizations with sufficient expertise to participate in international fora. Nonetheless, such coalitions and organizations could well emerge in the foreseeable future. Indeed, this process could be accelerated by enshrining the representation of (voting) non-governmental organizations from both camps—environment associations on the one hand and industry federations on the other—in the WEO statute.

The ILO formula is far from perfect, in particular given the higher degree of complexity in environmental policy compared to the more clear-cut 'business versus labour'-type of conflicts. And yet, the ILO provides a conceptual model along which ideas for an equitable participation of civil society in a WEO Assembly could be developed.

Environmental Protection versus Sustainable Development

Some have argued that the environment is too complex an issue to shape the mandate for a single organization. Calestous Juma (2000), for example, contends that a 'world environmental agency would need to cover every conceivable human activity'. Some participants in the debate have therefore proposed to create a 'world organization on sustainable development' instead of a 'world environment organization'.[11] Such a new organization is often conceptualized as a merger (and upgrade) of UNEP and UNDP (at least). I view this option as problematic: a merger of UNEP and UNDP would be a marriage of unequals that is likely to harm environmental interests in the long run without necessarily strengthening development goals.

First, UNDP and UNEP are unequal regarding their sheer size and resources. Taking into account the twelve-fold larger core budget of UNDP vis-à-vis the UNEP Environment Fund as well as a ratio of roughly four to one in professional staff, a merger of both programmes would come close to the dissolution of UNEP within the significantly larger UNDP. Theoretically, this could result either in a strengthening of environmental goals within the development community or in the slow degrading and watering down of environmental goals in a larger new, development-oriented agency. Key factors will be organizational culture and learning processes as well as leadership, both of which are important factors that help to explain organizational behaviour of international agencies (Leiteritz and Waever 2002; Siebenhüner 2003; Bauer 2004). Both UNEP and UNDP are marked by distinct organizational cultures tuned to the goals of the respective programmes. Given differences in size and resources, it is difficult to believe that the much smaller 'environmental' community will eventually prevail in changing the much larger 'development' community within an overall new organization. It seems certain that the strength and independence of environmental concerns will be weakened over time.

This is in the interest of neither North nor South, since, as a second argument, functional differentiation in governance systems between socio-economic development and environmental protection makes sense. Hardly any country has opted for the administrative merger of 'economic development' and 'environmental protection' as policy areas at the national level. Despite two decades of debate on sustainable development, we observe only very few examples of integrated 'ministries on sustainable development', with most countries maintaining the differentiation between

---

[11]   This section draws on, and has further been elaborated in, Biermann and Bauer, 2004b.

economic or development ministries, and environmental ministries. This national experience illustrates that environmental policy indeed can, and should, be addressed by one administrative unit. It is not clear why administrative functional differentiation should differ at the international level. Most international organizations and national ministries have clearly defined mandates for their respective policy areas, and it is theoretically not difficult to demarcate the responsibilities of a new international organization for the environment (see also Charnovitz, this volume). All this advises against the merger of UNEP and UNDP into one integrated programme or organization.

Third, UNEP and UNDP are unequal regarding their functions within their respective governance areas. UNEP has an important role in agenda-setting and knowledge-management, for example with a view to the initiation of new treaties, the organization of international diplomatic conferences, the training of national administrative and legal personnel, or the initiation, synthesis and dissemination of new knowledge, regarding both fundamental and applied environmental science. UNDP's core functions, on the other hand, are operational. It is mandated to generate and implement projects, with less regard to international standard-setting or knowledge-generation. A merger of UNDP and UNEP hence runs the risk that the different functions of UNEP will lose their influence within such a larger new agency.

If, on the other hand, a world organization on sustainable development would imply merely the upgrading of UNEP to an international organization with this name, while leaving other bodies—including UNDP or the World Bank—untouched, it is unclear what consequences the choice of the organization's name—'sustainable development' instead of 'environment'—would have. Opponents might rightfully complain that this would reduce the overarching concept of 'sustainable development' to what many Southern experts believe to be the Northern understanding: a new attractive yet deluding label for environmental protection (e.g. Agarwal et al. 1999). To the extent that sustainable development is understood as the established triad of socially, economically and ecologically sound development, one must object to a conceptualization of a world organization on sustainable development that addresses predominantly traditional environmental policy. In sum, a world organization on sustainable development would be either ill-advised if it implies the integration of UNEP and UNDP, or a misuse of a key concept of North-South relations if it merely implies giving a new name to an essentially environmental organization.

This does not, however, imply that a WEO should address environmental policy as unrelated to the larger quest for development. A WEO should aim at the preservation of environmental resources *within* the development process, not unlike the role of environmental ministries in developing countries. A WEO should not be seen in juxtaposition of environment *and* development, but rather within a framework of environment *for* development. The WEO constitution would hence have to encompass more than purely environmental rules, but address the development concerns of the South as well. Therefore, general principles such as the right to development, the sovereign right over natural resources within a country's jurisdiction or the principle of common but differentiated responsibilities and capabilities need to be integrated into the constitutive instrument of the world environment organization.

Southern Interests in a World Environment Organization

So far, only France, Germany and some other European nations, along with Brazil, Singapore and South Africa, have indicated their interest in supporting the idea of a world environment organization. Most developing countries have kept silent or voiced their scepticism about any attempt to strengthen global environmental governance without sufficient guarantees that their own right to development will not be negatively affected. Yet, a WEO might meet the interests *especially* of developing countries, for example by a more efficient and effective transfer of technology and financial aid to developing countries, and by a more efficient negotiation system that will increase the opportunities of (smaller) Southern nations to raise their voice in global fora (Biermann 2002b). It could also result in better coordination of Southern positions, which will in turn strengthen the collective bargaining power of developing countries.

Decision-making procedures based on North-South parity—i.e., veto rights for the South (and the North) as a group—could ensure that the WEO would not evolve into a new form of eco-colonialism, as many Southern actors and observers may fear: one solution could be a double-weighted majority system in the WEO Assembly, comparable to that of the

Montreal Protocol as amended in 1990 or of the Global Environment Facility as reformed in 1994.[12] In both institutions, decisions require the assent of two thirds of members that must include the simple majority of both developing and developed countries. This system of North-South parity in decision-making represents a new 'third path' between the one-country, one-vote formula of the UN General Assembly, which grants developing countries an in-built majority, and the one-dollar, one-vote system of the World Bank and the International Monetary Fund, which favours the interests of the major industrialized countries. Given that the concept of double-weighted voting has been developed in the environmental field, it seems to be an ideal basis for voting within the WEO Assembly.[13]

A world environment organization would also be likely to become the first UN specialized agency to reside in a developing country. Some Northern experts complain that Nairobi, the current seat of UNEP, would be a sub-optimal place for the headquarters of an efficient international bureaucracy, in particular since it is difficult to recruit highly qualified staff with families for positions in Nairobi. On the other hand, modern means of electronic communication have improved the situation substantially. Furthermore, the UN office in Nairobi remains the only UN headquarters in a developing country. It seems unlikely that UNEP—or a subsequent world environment organization—would be transferred to Geneva or New York: a

---

[12] Since 1994, GEF decisions require a two-third majority that must include sixty per cent of the states participating and sixty per cent of the financial contributions. In effect this North-South parity-based process grants both developing and industrialized countries effective veto rights.

[13] Admittedly, decision-making procedures based on North-South parity are problematic, especially when it comes to the question of which country belongs to which group (and who will decide this). Singapore, for example, is still seen as a developing country even though its national per capita income is higher than that of many industrialized countries. The Montreal Protocol uses an issue-based classification scheme that draws a line between developing countries that consume less than 300 gram chlorofluorocarbons per person and year, and all other countries (thus placing for example Kuwait in the group of industrialized countries) (Biermann 1997). Such issue-based classification, however, cannot be applied for a world environment organization meant to address all environmental issues. It seems that the second-best option remains the self-definition of states, as known from the UN General Assembly, with the expectation that certain developing countries—especially those that wish to join the OECD—will assume the respective additional responsibilities in international organizations, too.

new WEO would most likely hail if not from Nairobi, then at least from another developing country.

Clustering is Important but Only Part of the Solution

Clustering of some of the hundreds of multilateral environmental agreements has been proposed to address the apparent coordination problems in global environmental governance. Clustering could involve the relocation of treaty secretariats, including the streamlining of administrative services, as well as the co-scheduling of conferences of the parties to related conventions (for instance through back-to-back meetings); the clustering of environmental reporting and information generation and distribution, for example in uniform reports, scientific assessments and clearinghouses; or the synchronization of the meetings of treaty bodies (see UNEP 2001a, Hyvarinen 2001, Oberthür 2002 and von Moltke's contribution to this volume).

Such limited reform efforts, however, are no substitute for the upgrading of UNEP to a specialized agency and for the co-location and joint administration of convention secretariats within this body. Clustering can only be a first step for a larger reform effort. There are so many different levels of clusters for convention-related activities necessary that separate clusters at each of these levels would not solve the existing coordination problems, but could even exacerbate them. For example, convention-related efforts need to be clustered, at one level, according to the environmental medium that is to be protected. Examples would be those agreements that protect the atmosphere or those that protect the marine environment. Such form of clustering is required in particular regarding scientific research and assessment, since the behaviour, transportation and effects of greenhouse gases, ozone-depleting substances and persistent organic pollutants are the subject of similar and related scientific efforts and models.

At another level, however, convention-related efforts need to be clustered according to the human activity at the root of the problem, for example intensive agriculture, transportation, or industrial production. Yet such activity-based clusters would require a different cut. The climate convention, for example, would need to be clustered, for one, with the agreements affecting transportation (together with marine pollution treaties, for example); with agreements regulating industrial production (e.g., jointly with the agreements on ozone-depleting and persistent organic pollutants); with deforestation-related conventions, such as the biodiversity convention; and with soil-related conventions, like the desertification convention.

At a third level, clusters are needed to address common problems related to the environmental policy instrument chosen. One example would be a cluster of agreements that require restrictions in trade, for example trade in ozone-depleting substances, in endangered species, in persistent organic pollutants, in hazardous waste, or in genetically modified organisms. The practical implications could be joint programmes for the training of custom officials or joint information-sharing mechanisms. Another area of clustering would be capacity building in the South. Many environmental agreements have their own provisions on capacity building, partially even their own funding mechanism for these activities (e.g., the Montreal Protocol), without necessarily much coordination. This would, again, call for a different set of clusters. A fourth cut would be regional clusters.[14]

To cluster the environmental conventions according to all these levels could significantly increase the coordination deficits of the current system, instead of reducing them. If we consider the Montreal Protocol as an example, one would need to cluster it, first, with the climate convention and the Stockholm convention on persistent organic pollutants, based on the medium to be protected; second, with these and the Basel convention on hazardous waste and some ILO conventions in a cluster on industrial production; third, with the convention on trade in endangered species, the Basel convention and the Cartagena Protocol on Biosafety[15] and others in a

---

[14]   This cut slightly differs from deliberations in the UNEP group of environment ministers working on international environmental governance. Here, ministers have identified three possible methods of clustering: (1) at the thematic level (of issue-specific multilateral environmental agreements such as the chemical conventions, the biodiversity-related conventions, the regional seas conventions, and others); (2) at the functional level (functions such as reporting, capacity-building, trade policies, et cetera); and (3) at the regional level. See UNEP 2001a, para. 23 et seq.

[15]   On the trade-related aspects of the new Cartagena Protocol on Biosafety to the biodiversity convention, see Gupta 2000.

cluster on trade-related agreements; and fourth in a cluster of conventions that provide financial and technical assistance for developing countries. Such a system of multiple interrelated clusters is not a workable blueprint for a strengthened system of global environmental governance. While the idea of clustering is sound, it needs to be driven further towards an effective 'clustering of the clusters', that is, to the integration of these clusters into a comprehensive specialized UN environment agency.

4   Conclusion

This chapter has argued that the establishment of a world environment organization would improve coordination of global environmental governance, would pave the way for the elevation of environmental policies on the agenda of governments, international organizations and private organizations, assist in developing the capacities for environmental policy in African, Asian and Latin American countries, and strengthen the institutional environment for the negotiation of new conventions and action programmes as well as for the implementation and coordination of existing ones. Naturally, a world environment organization as outlined above cannot solve all problems of environmental degradation, but can only be a partial contribution. Yet this should not, I believe, result in a rejection of reform with the argument that improved governance is not 'a puzzle of administrative efficiency [but] a challenge of global justice' and that global cooperation is not 'a function of inappropriately designed organizations [but] a reflection of a fundamental absence of willingness on the part of states' (Najam 2003). It is, in most cases, both. Limited reform steps, such as improved regime design or more effective international agencies, will be influenced by the larger context of state power and global inequalities in the same way in which major reforms need to be implemented by solving myriad smaller questions. Students of global governance need to focus on the bigger questions, but also on the details—including the underlying organizational structure of global environmental governance.

Given the current state of global environmental governance, creating a world environment organization might appear unrealistic to some. The issue was hardly mentioned by governments in Johannesburg in 2002—which is true, however, for many issues discussed in the run-up to the Johannesburg summit. This might indicate either that reforms are impossible, or that the creation of a WEO, as a specific reform initiative, is an idea that will not find support by governments. Given the lack of concrete debate or action in Johannesburg, no clear answer can be given. Most major organizational and institutional reforms took their time: for example, a decade ago, the establishment of an International Criminal Court might have seemed unrealistic, too.

Of course, in the end the key issue is less organizational design and more the political will of governments regarding how much authority and resources they wish to grant international organizations, as rightly pointed out by Hans-Dieter Sohn (2001) or Adil Najam (this volume).[16] But analysts have to offer institutional formats through which decision-makers could best organize and coordinate their policies on international environmental issues. If governments chose to integrate various convention secretariats into a world environment organization, turf-battles would certainly emerge, as pointed out by Newell (2001) and Oberthür and Gehring (this volume). But these can be overcome if governments wish to do so, since it is governments who remain the principals of organizations, programmes and conferences of the parties.

The resistance to any streamlining effort by interested actors—including the heads of the various convention secretariats, who are likely to lose influence—is a practical problem rather than a theoretical obstacle to delineating a mandate for a world environment organization. A world environment organization would not solve all problems, neither of industrialized countries nor of developing countries. But it would be an important institutional step in humankind's efforts to both equitably and effectively manage planet Earth.

---

[16]   Another issue are the conditions for, and processes through which major reforms of the UN system in general and specific policy sectors can be achieved, e.g. regarding questions of the relevant actors, power distribution and interests. This set of questions is beyond the scope of this chapter, but is being currently analyzed by the MANUS research group ('Managers of Global Change: Effectiveness and Learning of International Environmental Organizations') within the international Global Governance Project GLOGOV.ORG. See Biermann and Bauer 2004a and forthcoming.

References

Agarwal, A., S. Narain and A. Sharma (1999), *Green Politics: Global Environmental Negotiations,* Centre for Science and Environment, New Delhi.

Andler, L. (forthcoming), *Splits or Bridge. The Case of the Global Environment Facility Secretariat,* Global Governance Working Paper, The Global Governance Project, Amsterdam, Berlin, Potsdam and Oldenburg [available at www.glogov.org].

Bauer, S. (2004), *Intergovernmental Treaty Secretariats and Policy Implementation: What Role for Bureaucracy?,* Paper presented at the Joint Sessions of the European Consortium for Political Research, Uppsala, Sweden, 13–18 April 2004.

Bauer, S. and F. Biermann (2005), 'The Debate on a World Environment Organization. An Introduction', in F. Biermann and S. Bauer (eds), *A World Environment Organization. Solution or Threat for International Environmental Governance?,* Ashgate, Aldershot, 1–23.

Biermann, F. (1997), 'Financing Environmental Policies in the South. Experiences from the Multilateral Ozone Fund', *International Environmental Affairs,* vol. 9 (3), 179–218.

Biermann, F. (2000), 'The Case for a World Environment Organization', *Environment,* vol. 20 (9), 22–31.

Biermann, F. (2001a), 'The Emerging Debate on the Need for a World Environment Organization: A Commentary', *Global Environmental Politics,* vol. 1 (1), 45–55.

Biermann, F. (2001b), 'The Rising Tide of Green Unilateralism in World Trade Law: Options for Reconciling the Emerging North-South Conflict', *Journal of World Trade,* vol. 35 (3), 421–448.

Biermann, F. (2002a), 'Institutions for Scientific Advice: Global Environmental Assessments and Their Influence in Developing Countries', *Global Governance,* vol. 8, 195–219.

Biermann, F. (2002b), 'Strengthening Green Global Governance in a Disparate World Society: Would a World Environment Organization Benefit the South?', *International Environmental Agreements: Politics, Law and Economics,* vol. 2, 297–315.

Biermann, F. and S. Bauer (2004a), 'Assessing the Effectiveness of Intergovernmental Organizations in International Environmental Politics', *Global Environmental Change. Human and Policy Dimensions,* vol. 14 (2), 189–193.

Biermann, F. and S. Bauer (2004b), *United Nations Development Programme (UNDP) and United Nations Environment Programme (UNEP),* Study prepared as contribution to the 2004 Report of the German Advisory Council on Global Change, Berlin.

Biermann, F. and S. Bauer (forthcoming), *Managers of Global Governance. Assessing and Explaining the Effectiveness of Intergovernmental Organisations,* Global Governance

Working Paper, The Global Governance Project, Amsterdam, Berlin, Potsdam and Oldenburg [available at www.glogov.org].

Biermann, F., R. Brohm and K. Dingwerth (2002) (eds), *Global Environmental Change and the Nation State: Proceedings of the 2001 Berlin Conference on the Human Dimensions of Global Environmental Change*, Potsdam Institute for Climate Impact Research, Potsdam [available at www.glogov.org].

Charnovitz, S. (2002), 'A World Environment Organization', *Columbia Journal of Environmental Law*, vol. 27 (2), 321–57.

Charnovitz, S. (2005), 'Toward a World Environment Organization: Reflections upon a Vital Debate', in F. Biermann and S. Bauer (eds), *A World Environment Organization. Solution or Threat for International Environmental Governance?*, Ashgate, Aldershot, 87–115.

Dingwerth, K. (forthcoming), 'The Democratic Legitimacy of Public-Private Rule-Making: What Can We Learn from the World Commission on Dams?', *Global Governance*.

Ehrmann, M. (1997), 'Die Globale Umweltfazilität (GEF)', *Zeitschrift für ausländisches öffentliches Recht und Völkerrecht*, vol. 57 (2–3), 565–614.

Esty, D.C. (1994), 'The Case for a Global Environmental Organization', in P.B. Kenen (ed.), *Managing the World Economy: Fifty Years After Bretton Woods*, Institute for International Economics, Washington, DC, 287–309.

Esty, D.C. (1996), 'Stepping Up to the Global Environmental Challenge', *Fordham Environmental Law Journal*, vol. 8 (1), 103–13.

Esty, D.C. and Ivanova, M. (2001), *Making Environmental Efforts Work: The Case for a Global Environmental Organization* (= Working Paper 2/01), Yale Center for Environmental Law and Policy, New Haven (Conn.).

Fairman, D. (1996), 'The Global Environment Facility: Haunted by the Shadow of the Future', in R.O. Keohane and M.A. Levy (eds), *Institutions for Environmental Aid. Pitfalls and Promise*, MIT Press, Cambridge (Mass.), 55–87.

Gupta, A. (2000), 'Governing Trade in Genetically Modified Organisms: The Cartagena Protocol on Biosafety', *Environment*, vol. 42 (4), 23–33.

Haas, P.M. (2002), *Science Policy for Multilateral Environmental Governance*, manuscript prepared for the United Nations University, University of Massachusetts, Amherst (Mass.).

Haas, P.M., N. Kanie and C.N. Murphy (forthcoming), 'Institutional Design and Institutional Reform for Sustainable Development', in P.M. Haas and N. Kanie (eds), *Emerging Forces in Environmental Governance*, United Nations University Press, Tokyo.

Hyvarinen, J. (2001), *Summary Notes on the Workshop on the Convention on Biological Diversity and International Environmental Governance, 8 October 2001*, Royal Institute of International Affairs, London [on file with author].

Jordan, A. (1994), 'Paying the Incremental Costs of Global Environmental Protection: The Evolving Role of GEF', *Environment*, vol. 36 (6), 12–36.

Juma, C. (2000), 'Stunting Green Progress', *Financial Times*, 6 July.

Kennan, G.F. (1970), 'To Prevent a World Wasteland: A Proposal', *Foreign Affairs*, vol. 48 (3), 401–413.

Keohane, R.O. and M.A. Levy (1996) (eds), *Institutions for Environmental Aid: Pitfalls and Promise*, Harvard University Press, Cambridge (Mass.).

Leiteritz, R.J. and C. Weaver (2002) *"Our Poverty is a World Full of Dreams": Organizational Culture and Change at the World Bank,* paper presented at the Critical Perspectives on Global Governance Workshop, at Schloss Amerang, Germany, 1–3 November.

Najam, A. (2003), 'The Case Against a New International Environmental Organization', *Global Governance*, vol. 9 (3), 367–84.

Najam, A. (2005), 'Neither Necessary, Nor Sufficient: Why Organizational Tinkering Will Not Improve Environmental Governance', in F. Biermann and S. Bauer (eds), *A World Environment Organization. Solution or Threat for International Environmental Governance?*, Ashgate, Aldershot, 235–256.

Newell, P. (2001), 'New Environmental Architectures and the Search for Effectiveness', *Global Environmental Politics*, vol. 1 (1), 35–44.

Oberthür, S. (2002), 'Clustering of Multilateral Environmental Agreements: Potentials and Limitations', *International Environmental Agreements: Politics, Law and Economics*, vol. 2, 317–40.

Oberthür, S. and T. Gehring (2005), 'Reforming International Environmental Governance: An Institutional Perspective on Proposals for a World Environment Organization', in F. Biermann and S. Bauer (eds), *A World Environment Organization. Solution or Threat for International Environmental Governance?*, Ashgate, Aldershot, 205–234.

Rajan, M.G. (1997), *Global Environmental Politics: India and the North–South Politics of Global Environmental Issues*, Oxford University Press, Delhi, Calcutta, Chennai, Mumbai.

Runge, C.F. (2001), 'A Global Environment Organization (GEO) and the World Trading System', *Journal of World Trade*, vol. 35 (4), 399–426.

Siebenhüner, B. (2003), *International Organisations as Learning Agents in the Emerging System of Global Governance. A Conceptual Framework,* Global Governance Working Paper no. 8, The Global Governance Project, Amsterdam, Berlin, Potsdam and Oldenburg [available at www.glogov.org].

Sohn, H.-D. (2001), 'Symbolic Solution?', *Down To Earth*, 15 December, 48–49.

South Centre (1996), *For a Strong and Democratic United Nations: A South Perspective on UN Reform*, South Centre, Geneva.

Trittin, J. (2002), 'The Role of the Nation State in International Environmental Policy: Speech by the German Federal Minister for the Environment, Nature Conservation and Nuclear Energy', in F. Biermann, R. Brohm and K. Dingwerth (eds), *Global Environmental Change and the Nation State: Proceedings of the 2001 Berlin Conference on the Human Dimensions of Global Environmental Change*, Potsdam Institute for Climate Impact Research, Potsdam, 10–13 [available at www.glogov.org].

UNEP [United Nations Environment Programme] (2001a), *Implementing the Clustering Strategy for Multilateral Environmental Agreements—A Framework.* Background paper for the fourth meeting of the Open-Ended Intergovernmental Group of Ministers or Their Representatives on International Environmental Governance, Montreal, 30 November–1 December 2001, UN Doc. UNEP/IGM/4/4 of 16 November.

UNEP (2001b), *International Environmental Governance: Report by the Executive Director,* prepared for the fourth meeting of the Open-Ended Intergovernmental Group of Ministers or Their Representatives on International Environmental Governance, Montreal, 30 November–1 December 2001, UN Doc. UNEP/IGM/4/3 of 16 November.

UNEP GC [Governing Council] (2001), *Proposals of the President of the United Nations Environment Programme Governing Council for Consideration by the Open-Ended Intergovernmental Group of Ministers or Their Representatives on International Environmental Governance* (presented at the fourth meeting of this group, Montreal, 30 November–1 December 2001), UN Doc. UNEP/IGM/UNEP/IGM/4/2 of 12 November.

von Moltke, K. (2005), 'Clustering International Environmental Agreements as an Alternative to a World Environment Organization', in F. Biermann and S. Bauer (eds), *A World Environment Organization. Solution or Threat for International Environmental Governance?*, Ashgate, Aldershot, 175–204.

Chapter 6

Generating Effective Global Environmental Governance: The North's Need for a World Environment Organization

John J. Kirton*

1   Introduction

As the 2002 Johannesburg World Summit on Sustainable Development (WSSD) fades into the past, the system of global environmental governance it codified, created or catalyzed hardly seems adequate to address the compounding ecological challenges the twenty-first century brings. After thirty years of heroic effort, from Stockholm 1972 through Rio 1992 to Johannesburg 2002, there was a feeling as Johannesburg concluded that much had been left undone and that little more would soon come. That sense has since spread. For despite some moves forward at the March 2003 Kyoto World Water Forum and the June 2003 Evian Summit of the Group of Eight (G8), the world is still

*   The author gratefully acknowledges the financial support of the Social Sciences and Humanities Research Council of Canada through the project on 'Strengthening Canada's Environmental Community Through International Regime Reform (EnviReform)', the comments of colleagues at a panel on 'Multilateral Environmental Agreements and Institutions: Making them Work in the Twenty-First Century World', at a conference on 'Canada @ the World', sponsored by the Policy Research Secretariat, Westin Hotel, Ottawa, 30 November 30-1 December 2000, the research assistance of Gina Stephens and Marilena Liguori, and the comments of Richard Ballhorn and Aaron Cosbey.

witnessing increasing, intersecting stresses on an integrated global ecosystem increasingly imperilled as economic globalization unfolds.

During the decades from Stockholm to Johannesburg, the international community had reacted to this compounding ecological stress in two major ways. The first had been an incremental accumulation of United Nations (UN)-based, subject-specific conferences and conventions to deal with the most visibly threatened individual media, pollutants and problems, punctuated by occasional 'summits' in 1972, 1992, 1997 and 2002 and intra-UN mechanisms such as the United Nations Environment Programme (UNEP) 1972 and the United Nations Commission on Sustainable Development (CSD) to provide coherence and central direction. The second way had been to build islands of intense institutionalized integrated cooperation at the restricted regional level, most notably in the European Union and, since 1994, in the North American Free Trade Agreement (NAFTA), its companion North American Agreement on Environmental Cooperation (NAAEC) and the latter's Commission for Environmental Cooperation (CEC).

The results of this dual approach are now clear. In an already rich and privileged Europe and North America, replete with strong national level regulation and capacity, growing economic integration and institutions have been accompanied by powerful ecological equivalents that are well integrated into the overall governance structure. At the global level, however, a poor and unprivileged South has been left unprotected, even as it lacks equivalent regional or national defences of its own. The world thus confronts ever more powerful processes and institutions of global economic governance, most notably the new World Trade Organization (WTO) and newly empowered International Monetary Fund (IMF) and World Bank, with no multilateral ecological institutions of comparable capacity and connection. The result is an ever-growing global gap between ever-stronger economic integration and institutions, on the one hand, and ever-weaker environmental quality and international organizational protectors on the other. The otherwise unprotected South may still be the first victim of this ecological-economic global governance gap (see Gupta, and Biermann, this volume). However, the globalization of ecological stress and vulnerability has now made the North, devoid of any outer institutional defences beyond its own borders, an equal victim as well. Its riches and regional institutions can no longer cope with the now globalized ecological assaults flowing in from the ungoverned South outside.

As the twenty-first century unfolds, the international community has reached the limits of the UN-based Stockholm-through-Rio-to-Johannesburg approach to protecting the global environment. For the resulting individualized, isolated, ad hoc multilateral agreements, accompanied by pockets of integrated regional institutions for part of the privileged North, can no longer cope with environmental challenges that are becoming more intense, interconnected, fully global and ever more devastating to the South and thus North as well. The time has thus come to construct a more comprehensive, integrated and ambitious architecture for generating global environmental governance for the twenty-first century, centred in a new World Environmental Organization (WEO) that is adequate to the task. The fact that Johannesburg 2003 failed to create stronger institutions for the environment, even as the WTO's Doha Development Agenda and the Free Trade of the Americas (FTAA) promised new waves of far reaching economic integration by 2005, suggest both the need for, and the urgency of, this task.

This chapter constructs the case for, and defines the resulting contours of, a desirable international architecture for global environmental governance for the twenty-first century. It further identifies why and how the countries of the North, through the Group of Eight (G8), might bring such an architecture and organization to life. It focuses on the intersection of the partial, often impotent environmental institutions with those powerful trade, investment and finance organizations at the heart of the economic globalization process. It argues that the increasing, interconnected global environmental stresses generated by economic globalization and its international organizations requires a system and centre of global environmental governance more comprehensive, coherent, capable and economically connected than is provided by the current system of individualized multilateral environmental agreements backed by a few robust regional organizations in the North.

Given its world leading environmental capabilities and vulnerabilities, the G8 has both an ability and an incentive to pioneer this new generation of environmental governance, centred on a powerful new WEO that both the G8 and the international community need. NAFTA offers a now proven model for the global architecture, that the G8's growth as an effective centre of global governance with a new outreach to the South provides a promising platform to produce such a WEO, and that the institutional failure of Johannesburg and prospective success of the WTO and some form of FTAA produce a powerful incentive to act.

This analysis thus makes several contributions to the rapidly growing debate over the need for, and shape of, a WEO. First, it is firmly grounded in the failure of global environmental, rather than economic or trade, governance, at a time when economic globalization has produced globalized ecological vulnerability that the UN-centric approach culminating in Johannesburg cannot offset (Runge 2001, 1994, von Moltke 2000). Secondly and specifically, from this broader vantage point, it claims that UNEP and the CSD, designed as weak institutions for a pre-economically globalized world, are now clearly architecturally and constitutionally inadequate for the new era, whatever their real or claimed successes in their own domains in their earlier years (Najam 2002). Thirdly, it denies that all failings of UN-based institutions can be reduced to the refusal of their member states to provide more money or political support to the same people, to do the same things, the same ways, and thus that the case for a WEO is based on mere administrative efficiency alone (Juma 2000).

Rather it claims that the principles, norms, rules and decision-making procedures, as well as the personnel and organizational culture of international institutions, have autonomous impacts on state and civil society behaviour, and that differences across these institutional attributes make a major difference in real world results. Fourthly, it argues that after a decade in operation, the model of the North American CEC and its NAFTA offer for a global WEO a proven architecture in which the values of the environment and the economy, and the countries of the North and South, have an equal and integrated place (Kirton 2003a, Rugman, Kirton and Soloway 1999). Fifthly, it suggests that after the failure of the Johannesburg Summit to create the needed new environmental organizations, that the Group of Eight (G8) major market democracies, with their proven environmental performance and resource raising capacity take the lead in bringing a WEO to life.

2   The Growing Globalized Ecological Challenge

Even those supportive of the institutional status quo in global environmental governance readily concede that the stresses on, and state of, the world's ecosystem and ecological carrying capacity are steadily getting worse (Najam 2002). However, their instinct for institutional incrementalism in response ignores three particular dynamics that twenty-first century globalization has now brought. The first is increasing stress on many of the world's critical environmental resources, which has now taken some beyond critical thresholds. The second is the interconnections among these media and pollutants revealed by unfolding scientific knowledge. The third is their growing global geographic nature, as physical ecological globalization unfolds faster than even economic globalization itself.

(a) Intensification. A wealth of credible public and private sector 'State of the Environment' reports reveal a global ecosystem under increasing stress, both overall and in critical components (UNEP 2000, 2002, World Bank 2000). To be sure, much uncertainty remains. Yet there is a wide range of areas where the global ecosystem is clearly under severe and increasing stress. These cover all the ambient media of air, water, land and living things, and pollutants and resource supports as well. The most acute include the depletion of the world fisheries and of forests. Further acute and visible threats include those to coral reefs, depletion of freshwater supplies in key regions such as China and the melting of polar ice under the impact of climate change. Recognition of the fact that environmental change can come slowly, cumulatively and invisibly, suggests that some still unmeasured and still-unrecognized losses lurk, especially in the domain of biodiversity (LePrestre 2002).

(b) Integration. These intensifying threats to critical individual ecological resources and media are accompanied by growing scientific evidence of their interconnections in an ultimately integrated global ecosystem. For example, the long known problem of acid rain in the domain of air has negative impacts on forests in the domain of land, and lakes in the domain of water, and diversity of the living things that exist therein. More recently, there has emerged an accepted political and scientific consensus, as the G8

leaders recognized as early as 1990 that an effective campaign to combat climate change must embrace all major sources and sinks, including forests and agricultural areas on the land, and even the oceans in the domain of water. Yet the UN-delivered climate change convention required over a decade before the sink value of forests and agricultural land was accepted.

(c) Globalization. This increasing stress on an integrated ecosystem is now rapidly moving, not merely from local to national in its geographic extent, but from regional to global as well, under the impact of contemporary economic globalization and development. As the Brundtland Commission recognized, increasing environmental stress from continuing population growth and industrialization creates pollution and compounding natural resource depletion in ever more distant locations, as the ecological footprint expands. These dynamics are now powerfully enhanced with the democratic-market revolution that is bringing so many emerging and transition economies, not only into improved environmental sensitivity and policies, but also into the rapid growth of the advanced industrial age.

The emerging result is the routinization of equal ecological inter-vulnerability. At the time of the Brundtland Report and the Rio Summit, it was relatively easy to accept the logic that the rich North should pay for the pollution that harmed them from their own rapacious resource development in the past, and compensate the still ecologically secure but impoverished South in advance for any environmental protection in which they might engage. Now it is clear that the South is the first victim of environmental degradation.

Equally importantly, an ecologically impoverished South now clearly causes an ecological vulnerability in the North that the national or even regional regime defences of even the most powerful rich countries are unable to control. While rich countries have more resources and thus more adaptive resilience, it matters little if the rich North maintains its favourable balance of relative invulnerability vis-à-vis the impoverished South, if even limited eco-invasions from the South overwhelm the geographic defences they construct across the North-South divide.

3   The Inadequate Global Environmental Governance Response

In the face of this intensifying, integrated, globalized ecological vulnerability, the international community offers an incomplete, inadequate and incoherent international institutional response. To be sure, there have been impressive developments at the multilateral and even global level in generating a vast array of legal agreements to govern many of the world's critical ecological resources. Similarly, there have been important innovations at the bilateral and regional level in creating international institutional and organizational capacity. Yet at the global level, there is an acute shortage of an adequate international institutional system and organization with the capability to deal in a comprehensive, coherent and compliance-inducing fashion with a global ecosystem under threat in a globalizing age. This lack of adequate global environmental institutional capacity is evident in the recent record of generating global and regional institutions in the environmental and economic field.

Incomplete Incrementalism: The Many Missing Institutions

The international community has had well over a century of experience in crafting multilateral environmental agreements (Meyer et al. 1997). Similarly, the emergence of multilateral environmental agreements with direct international economic implications, notably those with trade measures as implementing and enforcement devices, are also an inheritance of the nineteenth century, having begun in 1878. As of 2000, there were an estimated 175 to 200 multilateral environmental agreements, twenty of which have trade restricting provisions (Charnovitz 1996). In both domains, the growth of international legal instruments has by no means been a continuous or inevitable process. After the 1980-1958 dark ages, era of the the Brundtland Commission report in the mid-1980s through to UNCED at Rio in 1992, showed promise. But this era ended at Johannesburg, whose forward-looking institution-creating legacy was, unlike Rio, very sparse indeed (Chasek 2003). If the singular reliance on UN-based, once-a-decade multilateral summit-driven process continues, the world will face the coming decade of environmental challenges with the inherited institutional array (see Elliott, this volume, Johnson 2001, Dodds

2000). This inheritance includes the great global conferences and summits periodically held to drive global environmental governance forward. It further includes the actual multilateral environmental agreements concluded, either as a result of these summits and conferences or emerging outside and below.

This incrementally constructed inheritance means that major lacunae, and thus incomplete coverage, remain. Indeed, during the 1980s a series of global scientific reports identified ten principal environmental concerns for the coming decade (Speth 2002, 18-19). These were desertification, forests, species extinction, population growth, freshwater, overfishing, pesticides and POPs, climate change, acid rain and ozone. Almost a quarter of a century later, only three—ozone, pesticides and POPs, and desertification—have fully functioning global regimes in which all the major powers are involved. A further four—species, overfishing, climate and acid rain—have only regional regimes, or global regimes in which the United States remains absent or which have not come into force. Three— forests, population, freshwater—have essentially nothing at all.

The particular case of forestry is instructive. For here it is clearly not the lack of scientific understanding or absence of 'political will' on the part of the G7 major powers that has caused the global regime to remain unborn. Indeed, as long ago as 1990, United States President George Bush and all his G7 colleagues, at the Houston G7 Summit, agreed on the need for a global forest convention and formally committed to bringing it into being within two years. Yet over a decade later it is not in sight. Its absence is explained by the misplaced faith the G7 leaders subsequently invested in the UN and Rio process to produce the regime, both in 1992 and again in 1997. It produced in over a decade only an intergovernmental panel on forests as a discussion forum. In sharp contrast, where the G7 leaders did not wait for, or rely on, the UN but acted directly, as in the case of the G7 pilot project for tropical forests in Brazil, real ecologically enhancing action swiftly took place. More broadly, growing up to govern global forests has been an array of competing voluntary standards struggling to cope where the UN-grounded intergovernmental system has failed (Kirton and Trebilcock 2004).

Ineffective Institutionalization

In addition to the many missing media—or pollution—specific institutions and agreements, those that do exist are inadequate, impotent and ineffective in inducing their member governments, outsiders and civil society actors to invest and comply. Most generally, close observers of the existing repertoire point to three major defects (Johnson 2001). The first is the need to go beyond the consensus on issues, identification of priorities, adoption of principles, assembly of coherent and extensive action plans, and articulation of strategies to the actual implementation required to give these agreements practical effect. The second is the need for much greater capacity, including new financial transfers, to accomplish this task. The third is a much more effective participation of nongovernmental organizations, the private sector and civil society more generally in this process. The need was underscored by the former IMF Managing Director Michel Camdessus who, in his farewell address to the tenth United Nations Conference on Trade and Development (UNCTAD), proposed a major international effort to ensure effective implementation of the action plans of the United Nations conferences and summits of the 1990s (Camdessus 2000).

More specifically, those calling for a WEO offer a compelling and comprehensive catalogue of the functions where the existing global environmental institutions fail to meet the minimum test of organizational completeness and effectiveness. These start with the proper principles and norms and strong scientific capacity found to be the critical causes of effectiveness for international environmental regimes (Haas, Keohane and Levy 1993, Victor and Ausubel 2000: 139-141). They extend to funding and capacity creation, basic data gathering, dispute settlement and enforcement. Here the essential consensus concern is not that global environmental institutions lack the trade-sanctioning enforcement mechanisms that the trade system boasts, and thus look to 'green' the GATT/WTO and create a linked WEO as an enforcement mechanism or model for the diverse array of very structurally different and diverse problems in the environmental domain. Rather it is that virtually all global environmental institutions, with the possible exception of the ozone regime, fail to meet the minimum standards of good global governance on many of the basic functions an international institution can be called upon to perform. There is thus a strong consensus from independent scholars and close observers that international environmental institutions stand out as largely ineffective in many fields (Haas, Keohane and Levy 1993).

It is not simply that they have secretariats too small, or budgets too modest, or locations too distant, or national government members too controlling or any other singular sin that constitutes the fatal flaw. Indeed, the North American CEC shows how much can be achieved with very little by way of conventionally conceived resources, if the essential architectural design is correct (see below). Nor is it that in particular cases, the wrong configuration of institutional characteristics has been assembled to meet the particular problem structure at hand. It is rather a case of massive institutional failure writ large.

Nor can this comprehensive relative ineffectiveness be explained by the unusual uncertainty or physical complexity that exists in the environmental field, or the realist article of faith that all international institutions are ultimately doomed to fail. For in the cognate fields of health and weather, where similar physical uncertainty and complexity exist, the UN system has generated at times reasonably effective regimes. Nor does the cause of the observed ineffectiveness lie in the fact that individual environmental institutions were 'designed to fail'. Rather, it is the overall system of incrementally created, incomplete environmental institutions that has failed in its basic design.

Impotent Integrators: The Missing Centre From UNEP to the CSD

Central to the design of that system is the black hole at the centre, devoid of any force that makes the system complete, coordinated and coherent. This force could take the form of an institution that draws the existing pieces together, provides direction and support, and generates other components, as required, in functionally appropriate ways. Alternatively, it could take the form of a network that connects the parts in a self-organizing system and that progressively evolves to meet new needs and perform new functions as the situation demands. After over half a century since the 1945 creation, and after three decades since Stockholm, the UN-based system of incremental institutionalism, replete with the Economic and Social Council (ECOSOC), UNEP and the CSD, has neither form. Thus, while the UN system offers an array of theoretical advantages, such as an institutional nest which lowers transactions costs, a near universal membership that offers legitimacy and the capacity for grand geographic, functional and burden sharing bargains, it also contains some fundamental flaws that explain the systems' decidedly poor performance in practice over its first 59 years.

In the 1945 constitutional creation of the UN system, broad domains such as labour and health were brought together, given an identity as comprehensive, integrated issue areas, affirmed as values in the Charter, and thus integrated with other social and economic concerns in an overall architecture that ECOSOC could oversee. Then and to this day, the very existence, relevance to other concerns and thus value of the natural environment in its own right or for instrumental reasons has been entirely omitted from the Charter and thus from the constitutional core of the UN system. Equally absent have been generic principles, such as the precautionary principle, that have importance in the ecological domain. Moreover, the Charter simultaneously affirmed a wide range of values and principles whose realization (with the primary exception of human health), involve an increased stress on, or consumption of, unvalued ecological capital. Also consistent with the limited levels of scientific knowledge, industrialization and the pollution-resource depletion dynamic of 1945, no part of the UN institution itself or the functional agencies it inherited or created were assigned, let alone dedicated to fulfilling, ecological responsibilities.

Moreover in its key decision-making rules, beginning with the United Nations Security Council clauses, the UN system permanently entrenched provisions that gave what were to become the least environmentally sensitive principal powers—notably the Soviet Union/Russia and China—a predominant role, while confining what were to become more environmentally sensitive major powers—such as Canada, Germany, Italy and Japan—to a secondary rank. Such choices made it more difficult for the system to take up, as many bodies such as the World Resources Institute have suggested, the issues of environmental security, broadly defined in a swift, strong and sensitive way (Homer-Dixon 1993).

Despite subsequent Charter amendments and much institutional development, in over half a century, the environment has never come close to getting into the constitutional or institutional core. Such a complete absence in the Charter has made it more difficult to act subsequently on suggestions, which flourished on the 'Road to Rio 1992', such as converting the Trusteeship Council, which had lost its purpose, into the central high-level body in the UN system dedicated to environmental governance. With such a weak ideational and institutional foundation, environmental considerations were destined to remain a fragile and lagging add-on and afterthought to the far more powerful established core as the UN system evolved. The burst of activity at Stockholm in 1972 and such companion moves as the United Nations Convention on the Law of the Sea (UNCLOS), with its innovative incorporation of the custodianship principle in article 234, were easily stalled and silenced during the new cold war and the neo-liberal revolution of the first half of the 1980s, until the Brundtland Commission Report of the mid-1980s revived the process.

During the three decades defined by Stockholm and Rio, there was hope that the UN summit model, and injection of high level political will and global attention it brought, could produce the required great leaps forward, if in lumpy instalments and with unpredictable results (see Elliott, this volume). Yet the cumulative result is now clear. The environment's appearance as an integrated institutionalized, if third level, issue area, in the form of UNEP in 1972, was compromised by the subsequent creation of a second and rival coordinative centre, the CSD in 1992. It divided the centre's concern between environment and development and left the environment unconnected to the economic fields of most concern to the rich North. The new need to coordinate the many weak coordinators with different missions will require the single strong centre that the UN system lacks.

With these strikes against the global environment in the UN system, any partial measures such as 'clustering', bred largely by assumed political necessity more than architectural desirability, are unlikely to succeed. Indeed, the once promising candidate of climate change, as a prospective comprehensive intellectual cluster for the environment, has produced a Kyoto protocol that has not come into force seven years after its creation in 1997 and over a decade after its climate change convention 'nest' was proclaimed in 1992. Nor it is clear how, or if, the many now competing presumed coordinators—UNEP, CSD, Rio/Johannesburg and Millennium Summits—will or can be 'clustered' as well.

It is unlikely that a *primus inter pares* will emerge from any of them. For the institutional failures of both UNEP and the CSD are well known. UNEP was created as, and over three decades later remains, a mere programme rather than a full-fledged functional agency of the UN, sporting a modest budget of 60 million US dollar per year. It receives only 5 per cent of its budget from the UN and its regular assessments, leaving it dependent on voluntary contributions, largely from seven donor countries, for the remaining 95 per cent. It is headquartered in distant Nairobi, with often scarce electricity, communications, water, personal safety and thus top-tier personnel. It is far removed from Geneva, New York or Washington, and from the tiny convention-specific secretariats that emerged from Rio and were located in Bonn (climate change) and Montreal (biological diversity) (Dodds 2000, LePrestre 2002). With such fragmentation and fragility in both legal powers and organizational capacity, it is understandable that these and similar institutions have had difficulty in functioning as effective international environmental regimes (Bernauer 1995, Haas, Keohane and Levy 1993, Litfin 1997, Sprinz and Vaahtoranta 1994).

Two decades later, the 1992 Rio conference did relatively little to redress the 1945 Charter imbalance as the Declaration on Environment and Development was far different than the genuine Earth Charter that many participants desired. Institutionally, Rio's major legacy was a mid-level institutional add-on, CSD, established as a functional body under the authority of ECOSOC. At CSD the representatives of the 53 states elected by the Council for up to three year terms meet once a year for two or three weeks (Dodds 2000). Equally limited in authority and organizational stature, before their 2001 demise, were the Inter-Agency Committee on Sustainable Development (IACSD), established as a subsidiary body of the UN Administrative Committee on Coordination (ACC), chaired by an Undersecretary General and composed of senior level officials from nine members of the ACC.

What hopes remained for this historic UN-based approach were brutally ended in 2002, with the World Food Summit in Rome in June and the World Summit on Sustainable Development in Johannesburg in September. Johannesburg did nothing of consequence to add any of the missing environment regimes, or strengthen or solve the competing coordinative relationship between UNEP and the CSD. Rather it turned away from any emphasis on intergovernmental institution building toward private partnerships. The summit-driven UN-centred system thus offers little prospect of generating the needed completeness and coherence in the years ahead.

4   The Growing Gap with Deepening Institutionalized Economic Globalization

With this dismal outlook for UN institutional development during the coming decade, the world's ecosystem faces an ever growing gap with the ongoing deepening economic globalization and the more powerful global economic organizations it brings. This gap is much greater than the conventional focus on trade-environment issues alone would suggest.

At the centre of the global economic governance system, in the field of finance, stands the world's most powerful international organization, the IMF. Created even before the UN itself, with sweeping powers to govern global finance, it was born with no environmental mandate or awareness. It has acquired none of consequence in the decades since, even as its power has grown during the past decade under the impact of financial globalization and the global financial crisis this has brought (Kaiser, Kirton and Daniels 2000, Kirton and Richardson 1995). The IMF, through its involvement in the Highly Indebted Poor Countries programme, structural adjustment and other activities, has far more impact than the WTO on the trade, investment, development and economic life of many developing countries. The IMF does lend in response to natural disasters such as floods and hurricanes, but only on a reactive basis and for a very limited array of classic environmental threats. Its ad hoc support programmes during the global financial crisis of 1997-99, for such ecologically critical countries as Brazil, devoted no attention to environmental values amidst the many quite detailed micro and structural conditions it imposed. Its Poverty Reduction and Growth Facility, which offers very cheap credit, focuses on health and primary education rather than core environmental concerns. Nor does the IMF include any serious sustainability analysis before it decide whether and how to make a loan.

The IMF's Bretton Woods sister, the World Bank Group, has gone a considerable way in recent years to transform itself from a singular economic development to an ecologically sustainable development institution. Yet environmental values remain a secondary add on. Moreover, the strength of the World Bank further unbalances the bias of institutionalized global governance in the direction of the development values favoured by the collective 'South', rather than the environmental values often pioneered in the North.

In the field of foreign direct investment, there is an amalgam of component regimes grounded in the Organization for Economic Cooperation and Development (OECD), UNCTAD and WTO codes. Here environmental considerations are effectively absent. The OECD does offer a membership with considerable global diversity in geographic location and level of development, a consensus and analytic-scientifically oriented culture, a proven track record of environment-economy innovation and an institutional structure that has allowed for meaningful civil society participation. Yet the OECD, as its name suggests, remains at its core an economic institution, in which civil society representatives of the environmental community have no comparable place to that accorded their business and organized labour colleagues. Its tendency is to privilege the ideology of 'economism', as its seminal framework for assessing the environmental effects of trade shows. It ultimately cannot be converted to serve as the substitute environmental organization even for the North alone.

In the field of trade, the great leap forward came in 1995, when the General Agreements on Tariffs and Trade (GATT) was transformed into a much more powerful WTO, now governing a single undertaking that encompasses a wide array of non-trade, if trade-related fields. Yet the now Southern dominated organization has retained its traditional aversion to incorporating environmental values, especially those privileged by the North. The most recent Doha development mandate, launching a negotiating round of further liberalization due to be completed at the end of 2005, did mark major advances for the incorporation of environmental values. Yet even if the Doha development agenda is completed on time with its environmental provisions largely intact and actually delivering the promised results, the central thrust, as with the World Bank, is development rather than the environment itself.

5   The NAFTA-CEC Model

While the incremental inheritance of UN-based environmental institutions with global reach and relevance lacks comprehensive coverage, effective implementation capacity, collective coherence and the strength necessary to counter the growing institutions of economic globalization and development, the decade of the 1990s has seen much more balanced and integrated institutional progress in much of the rich North. In most respects, Europe through the European Union (EU) has been in the lead. Asia, either as a self-contained region or in its trans-Pacific extension through the Asia-Pacific Economic Cooperation (APEC) forum has lagged (Rugman and Soloway 1997). It is easy to understand why leadership has come from the European Union, for its emerging 'United States of Europe' has the same scale as North America, more dense transborder interdependencies, and members drawn entirely from the rich, democratic North. In contrast, Asia, especially with the Pacific component, offers much greater geographic scale, fewer physical transborder interdependencies with political visibility, and dominance by a diverse array of open and closed developing countries within.

In the middle, as the relevant model for global institution-building, stands North America. It offers in its NAFTA, accompanying NAAEC and component CEC, the world's first and only integrated full free trade and environmental agreement and institution in which countries from the rich north and once poor south come together as equals (Runge 2001, 1994). A decade after the regime came into formal effect, it has a proven performance as an effective environmental and linked environment-economy regime, grounded in an international organization with few fully owned resources (including a budget of only 9 million US dollar a year), where the United States participates but as a minority voting member, and where two US administrations—Bill Clinton's Democrats and George W. Bush's Republicans—have offered consistent support.

Although Canada and the United States had created a host of environmental agreements and institutions throughout the twentieth century, most notably the Boundary Waters Treaty and International Joint Commission (Spencer, Kirton and Nossal 1981), the CEC was a revolutionary creation in several respects (Rugman, Kirton and Soloway 1999, Kirton and Fernandez de Castro 1997). It was a regime that embraced as equals countries of North and South. It recognized the existence of, and need to manage, the ecological interdependencies on a wide and disparate regional rather than merely transboundary scale. Through its Joint Public Advisory Committee and National Advisory Committees it gave civil society an integral place in the governance of the institution as a whole. The regime allowed existing treaties, agreements and institutions to continue, but created a single new comprehensive centre, rather than slivers of incrementally created issue specific agreements or institutions, to manage the environmental challenges and temper the economic liberalization that the region faced. It understood the centrality of directly integrating the environmental institution into the new regimes for trade and investment liberalization. And it created North America's first real regional organization to manage the process.

It now is a regime where, uniquely in the global community, the environmental body is organizationally stronger than that of its trade-investment-finance counterparts. Indeed, from an international organizational perspective, in North America the environment comes first. Together the architecture of the NAFTA-NAAEC-CEC regime meets all of what might be termed 'the von Moltke conditions' for what a WEO must be like (von Moltke 2000).[1] In particular, it contains a substantial cumulative North-to-South capacity building component, through the training and technical assistance aspects of its trilateral programme, through the parallel Border Environmental Cooperation Commission (BECC) and North American Development Bank (NADBANK) on the critical United States-Mexican border and through the North American Fund for Environmental Cooperation which the CEC added soon after the organization began its work.

After ten years in operation, the CEC has, in its programmes of stand-alone environmental cooperation, substantially met the mandate it

---

[1] This is hardly surprising as Konrad von Moltke was the intellectual founding father of the CEC and always argued a North American Commission for the Environment, or NACE as it was first labelled, was necessary for environmental reasons alone, even if the economic NAFTA enhanced the need and provided the necessary political catalyst for the CEC to be born.

was given, if not the much larger expectations and potential that surrounded its birth (Esty et al. 2000). It and its NAFTA sister have been less, but still somewhat, successful in fulfilling the legal obligations, encoded in both the core NAFTA trade-investment agreement and the parallel NAAEC, to bring ecological considerations to bear on the trade and investment liberalization which NAFTA unleashed. Yet across the full array of functions, from state-of-the-art environment monitoring through to dispute settlement and enforcement, there is solid evidence that the CEC-NAAEC-NAFTA model works. Indeed, it works for both the environment and the trade communities (Rugman, Kirton and Soloway 1999). It works for each of the three member countries as well (Kirton 2003a, 2002a).

6   Going Global with a WEO

With the North American regional regime as an appropriate and proven model for global environmental governance, the next question is whether the rich Northern countries, led by the G8 democratic powers, will need to, and want to, build a global equivalent as part of their international environmental diplomacy repertoire. The evidence suggests they will.

At the national level, all G8 and OECD governments, led by the United States, have, since the late 1960s, created and continued a comprehensively oriented environment ministry or agency, even as they have retained existing strong sectoral departments to deal with particular environmental resources such as fisheries, forestry and agriculture. In over three decades not a single one of these governments have found it preferable to abolish these central environmental bodies in favour of a fragmented approach. They thus all find it normal, natural and necessary to have a central environmental organization. It is important to note that the first call for a WEO dates from this time when national environmental departments were being created (Kennan 1970). It came at a time long before the GATT/WTO trade system ran into trouble, before the creation of UNEP and as a natural global twin of central environmental ministries being created in Northern countries at home.

Thus empowered nationally, these OECD and especially G8 governments have pursued their global environmental diplomacy through several approaches. One is a strategy of national closure—defending the national ecosystem at the border—through autonomous national regulatory and inspection regimes. Yet as Europeans and now North Americans recognize, there are many fewer opportunities for effective border defences in a globalizing age, notwithstanding new national investments in agricultural, hazardous waste and other national inspection regimes. This has remained true even after the shock of 11 September 2001 brought a major thrust toward border closure in the United States. It is true not only for compacted continental Europe but also for the ecological superpowers of the United States, Russia and Canada—transcontinental countries with a vast array of open borders and long coastlines, located at the intersection of three major oceanic ecosystems and containing the fragile Arctic environment.

Unilateralism, or forceful action beyond the national border, has a residual relevance, even for those countries that consider themselves modest middle powers devoted to the UN-based multilateralist cause. Even, the G8's long weakest member, Canada, has recurrently practiced well-targeted, effective environment-economy unilateralism, as with the 1970 Arctic Waters Pollution Prevention Act, the use of force in the spring of 1995 against Spanish overfishing on Canada's east coast and in its threat of trade restrictions to induce South Korea to reduce its fishing off Canada's east coast. Yet these actions have been not an end in themselves, but a temporary tactic in creating innovative multilateral regimes that the UN system itself could not produce. The results, from article 234 of the United Nations Convention on the Law of the Sea to the 1995 United Nations Agreement on Straddling and Highly Migratory Fish Stocks, are landmark achievements that few Northern or Southern countries would wish to repudiate today.

Another approach, also practised in the face of the UN's broad multilateral stalemate, is the fostering of voluntary private sector-driven standardization regimes (Kirton and Trebilcock 2004). This approach has made a contribution, from the International Standards Organization (ISO) and its 14000 systems standards, to the Codex Alimentarius, to the efforts in forestry where the formal intergovernmental process has failed. Yet such standards can be slow to create, as they are based on consensus and difficult to enforce. They can inhibit a vibrant regulatory race to the top as well as the much feared, if seldom seen, regulatory race to the bottom. They can create costly competing standards regimes. Nor do process standards of the ISO variety clearly and quickly deliver demonstrated improvement in environmental performance. Above all, they are always vulnerable to defection, leading some industries to look, after their initial appeal, to have them entrenched in regulatory action by governments with authority and the full force of 'hard' law. There are thus sound reasons to believe that global corporations would support a new WEO.

7   Generating a G8-Centric WEO

There are thus good grounds for constructing a hard law, intergovernmental regime at the global level, centred in a strong, comprehensively oriented environmental organization, of the sort that all G8 and OECD governments have long found necessary at the national level and, in the cases of North America and Europe, at the regional level as well. The remaining question is how this organization should be created, especially in a post-Johannesburg era where the UN-based summit model has so strikingly failed (Newell 2002).

One possible path is to multilateralize the CEC model by having the North American regime serve as the platform and the growth pole that an ever expanding array of other countries from around the world could join. Both the United States and Canada are rapidly creating bilateral free trade agreements, containing NAFTA-like trade-environmental provisions, with developing countries in the Western hemisphere and in the Middle East, Africa and Asia as well. Here the next step would be to continue this process and have the CEC itself, with expanded resources and adjusted membership provisions, serve as the single environmental organization serving all these North American grounded individual 'hub-and-spoke' regimes. The first step was envisaged by, and built into, the initial NAFTA architecture, with its provisions for NAFTA accession that were not geographically limited but open to any country in the world. However, relying on this approach would still take time, bypass many of the major emerging ecological powers from the developing world at the start, and leave the pace and path of the environmental organization dependent upon that of the free trade process. Far more than NAFTA, it would represent a 'trade first-environment second' approach.

A second possible path is to use, as NAFTA did at the regional level in 1992, the 'big bang' of more multilateral trade liberalization as the necessary spur to bring a linked WEO to life. Here the most proximate prospective action-forcing deadline is the end of 2005. Then both the regional FTAA and the more multilateral Doha development agenda of the WTO are destined to conclude. In the case of the WTO, the environmentally friendly Doha Development negotiating mandate, and the presence and weight of the Europeans, Canadians and Japanese, make it possible that an environmentally oriented result, with a link to a new WEO, might result. However, few predict that the negotiation will conclude by its intended date, which would make its conclusion twice as fast as previous rounds have been. More likely to arrive in time in some form is the FTAA, with a constituency already substantially attached through bilateral free trade agreements to the North American model (Segger et al. 1999). Yet is seems unlikely that either the United States or Brazil, the key drivers at the negotiating endgame, will champion a WEO or a NAFTA-like Free Trade Agreement of the Americas trade-environment regime. Here the deadweight of an environmentally unfriendly and development-oriented Organization of American States and Inter-American Development Bank in the hemisphere also crowd out the space in which a new regional organization dedicated to the environment could be born.[2]

The third and preferable path is to rely on the G8 major power democracies to bring a new WEO to life. Indeed, it was through the leadership of a small group of the ten advanced industrial, urban and maritime powers, from both the communist and non-communist blocs, that a WEO was seen to spring to life in the very first call for such a body to be born (Kennan 1970: 410). With these powers now already collectively organized into the G8 summit system, with a now democratic Russia now as a full member and with the European Union representing an expanding

---

2   It is important to note that close observers of the ecological needs of Latin America and the Caribbean support the creation of a WEO rather than a regional or hemispheric environmental organization (Schatan 2002). More broadly, the 2002 creation of the African Union, with many new regional institutions but none for the environment, suggests that a regional substitute or way station for a WEO will not work.

array of other relevant states, it is much easier for such plurilateral leadership from the powerful to flow. Moreover, the proven record of the G8 in the environment field makes it feasible for this act of creation to come (Kirton 1999).

The G8 has dealt with environmental issues at every one of its annual summits since the institution was founded in 1975. During this time it has dealt with virtually every environmental issue, and with environment-economy, environment-health and environment-security ones as well. As a flexible, leaders-driven institution, the G8 is free to deal with any new or old environmental issue and linkage, combine them in new ways to establish innovative principles and norms, and produce hard decisional commitments from which national compliance flows. Both overall and in the environment field the G8 has produced a large number of often ambitious and significant commitments. Its member governments, from the most economically powerful United States to the least economically powerful Canada and now Russia, have complied with these commitments to a substantial degree (Kokotsis and Daniels 1999). Moreover, in sharp contrast to the UN system, the G8 has found it relatively easy to mobilize massive sums of money from its members for environmentally related purposes. During the past decade this list includes supporting economic reform in Russia, the states of the former Soviet Union, and central and eastern Europe, ensuring nuclear safety in Ukraine, relieving the debt of the world's poorest countries, combating infectious disease, reducing poverty in Africa and safely eliminating weapons of mass destruction. During the post-Rio decade, the world's major democratic powers have consistently and readily chosen to invest in and through the G8, rather than the UN. In short, if one wants an environmental organization where rhetoric moves rapidly into real commitments, and implementing action backed by massive resources, one would starting by building it in the G8 and not the UN.

Moreover, during the post-Rio decade of intensifying economic globalization, the G8 has proven to be institutionally generative on an ever expanding scale. Since 1992, the G8 has produced a new galaxy of stand-alone ministerial institutions, with its ministerial forum for the environment, created from 1992 to 1994, in the lead. More recently, at the Okinawa Summit in 2000, the G8 moved to pioneer, with the Renewable Energies Task Force, official level bodies, in emerging areas that built in actors beyond the G8 and civil society participants from the start. More recently, the G8 has incorporated in its annual summit a broad array of developing countries, from all locations and strata of the 'South' and from international organizations, giving the G8 a more representative, legitimate and effective global reach.

Within the G8, there has long been an emphasis on defining global environmental governance, and environment-economy governance, without creating heavy, hard law multilateral institutions which could constrain the flexibility of democratic leaders to adjust in a timely, flexible fashion to the needs of a rapidly globalizing world. But during the 1990's, the G7/8 found it easier to reliably forge the badly needed effective, balanced trade-labour linkages, where the hard law multilateral International Labour Organization (ILO) existed, than trade-environment ones, where there was no WEO to balance and integrate the core values of the WTO, IMF and World Bank (Kirton 2002b). At the same time, the G7/8 has long produced new, effective institutions and regimes that began with only G8 and other rich northern G7/8 countries as members, but that have rapidly radiated outward to become a genuinely global regime. The recent G7-centered Global Health Security Initiative, the current United States-led, G8-centric effort to create a comprehensive, integrated earth observation network and international partnership for a hydrogen economy, and Canada's initiative to establish a leaders-level G20 all show this impulse at work.[3] In short the G8 can generate the WEO that the twenty-first century world needs.

Over the coming years the G8 will be hosted by environmentally friendly governments that could well be willing to use their prerogatives as G8 host to take up the WEO-creation cause. This cadence can commence with the United Kingdom in 2005, continue through Russia, hosting for the first time in 2006 and culminate with Germany in 2007. While Russia might seem to some as a very weak link, its position as the possessor of world-leading ecological and energy capabilities, its second class status in the economic institutions that the 1944/1945 settlement produced and its

---

[3]   For another example of this process, in regard to forests, see Victor and Ausubel (2000: 141-142). See also Bryner (1997: 196).

recent participation in the G8's environmentally enhancing Global Partnership on Weapons of Mass Destruction give good grounds for placing it in the G8's 'green' club (Kirton 2003b).

8   Conclusion

It is now over a third of a century since the great US geopolitical thinker George Kennan issued his seminal 1970 call for an 'International Environmental Agency' possessing a comprehensive array of functions, collaborating with scholars, scientists and experts, controlled by true international servants, and enjoying 'great prestige, great authority and active support from centres of influence within the world's most powerful industrial and maritime nations' (Kennan 1970: 409). Kennan's analysis has proven incorrect only in its implication that the creation of such a single commanding organization was inevitable. But everywhere else its prescient analysis and proposed architecture has remained true to this day.

The four fundamental changes that have come since Kennan's call give his case, concept and roadmap for creation added force. The first change is an intensifying economic globalization that generates unbearable stress for the global environment, produces an equal ecological intervulnerability between North and South, and yields development-oriented international economic organizations far more powerful than those created over half a century ago. The second is the growing gap between these economic institutions and values on the one hand, and their ecological equivalents on the other, as the basic approach of incremental, individual, isolated UN-based institution-building for the environment has now clearly proven unable to keep up and cope. The third is the failure, after Johannesburg 2002, of relying on once-a-decade UN summits to provide the impetus for periodic great leaps forward in the international environmental institution building task. And the fourth is the emergence and now proven effectiveness of the G8 system as an increasingly inclusive centre of global governance.[4] The G8 cannot meet the ecological needs of the North, nor substitute for a hard law, properly multilateral WEO, any more than the OECD could in the past. But the G8, especially if reinforced with a leader-level G20, can serve as the platform on which and through which the long needed WEO can finally be built.

References

Bernauer, T. (1995), 'International Institutions and the Environment', *International Organization*, vol. 49 (Spring), 351–77.

Biermann, F. (2005), 'The Rationale for a World Environment Organization', in F. Biermann and S. Bauer (eds), *A World Environment Organization. Solution or Threat for Effective International Environmental Governance?*, Ashgate, Aldershot, 117–144.

Bryner, G. (1997), *From Promises to Performance: Achieving Global Environmental Goals*, W.W. Norton, New York.

Camdessus, M. (2000), 'Development and Poverty Reduction: A Multilateral Approach', Address at the Tenth United Nations Conference on Trade and Development, Bangkok, 13 February.

Charnovitz, S. (1996), 'Trade Measures and the Design of International Regimes', *Journal of Environment and Development* 5 (2), 168–96. Also in A. Rugman and J. Kirton with J. Soloway (1998) (eds), *Trade and the Environment: Economic, Legal and Policy Perspectives*, Edward Elgar, Cheltenham, 444–72.

Charnovitz, S. (2005), 'Toward a World Environment Organization: Reflections upon a Vital Debate', in F. Biermann and S. Bauer (eds), *A World Environment Organization. Solution or Threat for Effective International Environmental Governance?*, Ashgate, Aldershot, 87–115.

Chasek, P. (2003), 'The Negotiating System of Environment and Development: A Ten-Year Review', Paper presented at the International Studies Association 44th Annual Conference, Portland, 25 February–1 March 2003.

Dodds, F. (2000), 'Reforming the International Institutions', in F. Dodds (eds), *Earth Summit 2002: A New Deal*, Earthscan Publications, London, 290–314.

---

4   But see Charnovitz in this volume.

Elliott, L. (2004), 'The United Nations' Record on Environmental Governance: An Assessment', in F. Biermann and S. Bauer (eds), *A World Environment Organization. Solution or Threat for Effective International Environmental Governance?*, Ashgate, Aldershot, 27–56.

Esty, D. et al. (2000), *NAFTA and the Environment: Seven Years On*, Institute for International Economics, Washington, DC.

Gupta, J. (2005), 'Global Environmental Governance: Challenges for the South from a Theoretical Perspective', in F. Biermann and S. Bauer (eds), *A World Environment Organization. Solution or Threat for Effective International Environmental Governance?*, Ashgate, Aldershot, 61–85.

Haas, P., R.O. Keohane and M. Levy (1993) (eds), *Institutions for the Earth: Sources of Effective International Environmental Protection*, MIT Press, London.

Homer-Dixon, T. (1993), *Physical Dimensions of Global Change. Global Accord: Environmental Challenges and International Responses*, MIT Press, Cambridge (Mass.).

Johnson, P.M. (2001), 'Creating Sustainable Global Governance', in J. Kirton, J. Daniels and A. Freytag (eds), *Guiding Global Order: G8 Governance in the Twenty-First Century*, Ashgate, Aldershot, 245–80.

Juma, C. (2000), 'The Perils of Centralizing Global Environmental Governance', *Environment Matters*, vol. 6 (12), 13–15.

Kaiser, K., J. Kirton and J. Daniels (2000) (eds), *Shaping a New International Financial System: Challenges of Governance in a Globalizing World*, Ashgate, Aldershot.

Kennan, G. (1970), 'To Prevent a World Wasteland: A Proposal', *Foreign Affairs* 48 (April), 401–13.

Kirton, J. (1999), 'Explaining G8 Effectiveness', in M. Hodges, J. Kirton and J. Daniels (eds), *The G8's Role in the New Millennium*, Ashgate, Aldershot, 45–68.

Kirton, J. (2002a), *International Institutions, Sustainability Knowledge and Policy Change: The North American Experience*, Paper presented at the 2002 Berlin Conference on the Human Dimensions of Global Environmental Change, Berlin, December 6–7.

Kirton, J. (2002b), 'Embedded Ecologism and Institutional Inequality: Linking Trade, Environment and Social Cohesion in the G8', in J. Kirton and V. Maclaren (eds), *Linking Trade, Environment and Social Cohesion: North American Experiences, Global Challenges*, Ashgate, Aldershot, 45–72.

Kirton, J. (2003a), 'NAFTA's Trade-Environment Regime and its Commission for Environmental Cooperation: Contributions and Challenges Ten Years On', *Canadian Journal of Regional Science*, vol. 25 (2), 135–63.

Kirton, J. (2003b), 'Governing Globalization: The G8's Contribution for the Twenty-First Century', in V. Razumovsky (ed.), *Russia Within the Group of Eight*, Institute for Applied International Research, Moscow.

Kirton, J. and S. Richardson (1995) (eds), *The Halifax Summit, Sustainable Development and International Institutional Reform*, National Roundtable on the Environment and the Economy, Ottawa, 133.

Kirton, J. and R. Fernandez de Castro (1997), *NAFTA's Institutions: The Environmental Potential and Performance of the NAFTA Free Trade Commission and Related Bodies*, Commission for Environmental Cooperation, Montreal.

Kirton, J. and M. Trebilcock (2004), *Hard Choices, Soft Law: Voluntary Standards for Global Trade, Environment and Social Governance*, Ashgate, Aldershot.

Kokotsis, E. and J. Daniels (1999), 'G8 Summits and Compliance', in M. Hodges, J. Kirton and J. Daniels (eds), *The G8's Role in the New Millennium*, Ashgate, Aldershot, 75–94.

LePrestre, P. (2002) (ed.), *Governing Global Biodiversity: The Evolution and Implementation of the Convention on Biological Diversity*, Ashgate, Aldershot.

Litfin, K. (1997), 'Sovereignty in World Ecopolitics', *Mershon International Studies Review*, 167–204.

Meyer, J.W., D.J. Frank, A. Hironaka, E. Schofer and N.B. Tuma (1997), 'The Structuring of a World Environment Regime, 1870–1990', *International Organization*, vol. 51 (4), 623–51.

Najam, A. (2002), 'The Case against GEO, WEO, or Whatever-else-EO', in D. Brack and J. Hyvarinen (eds), *Global Environmental Institutions: Perspectives on Reform*, Royal Institute of International Affairs, London.

Newell, P. (2002), 'A World Environment Organisation: The Wrong Solution to the Wrong Problem', *World Economy*, vol. 25 (5), 659–71.

Rugman, A. and J. Soloway (1997), 'An Environmental Agenda for APEC: Lessons from NAFTA', *International Executive*, vol. 39 (6), 735–44.

Rugman, A., J. Kirton and J. Soloway (1999), *Environmental Regulations and Corporate Strategy: A NAFTA Perspective*, Oxford University Press, Oxford.

Runge, F. (2001), 'A Global Environmental Organization (GEO) and the World Trading System', *Journal of World Trade*, vol. 35 (4), 399–426.

Runge, F., with F. Ortalo-Magne and P. Vande Kamp (1994), *Freer Trade, Protected Environment: Balancing Trade Liberalization and Environmental Interests*, Council on Foreign Relations, New York.

Schatan, C. (2002), 'World Environmental Organization: A Latin American Perspective', *World Economy*, vol. 25 (5), 673–84.

Segger, M.-C., M.B. Muños, P.R. Meirles, J.Z. Taurel and V. Paul (1999), *Trade Rules and Sustainability in the Americas*, International Institute for Sustainable Development, Winnipeg.

Spencer, R., J. Kirton and K.R. Nossal (1981) (eds), *The International Joint Commission Seventy Years On*, Centre for International Studies, University of Toronto, Toronto.

Speth, G. (2002), 'The Global Environmental Agenda: Origins and Prospects', in D. Esty and M. Ivanova (eds), *Global Environmental Governance: Options and Opportunities*, Yale School of Forestry and Environmental Studies, New Haven (Conn.), 11–30.

Sprinz, D. and T. Vaahtoranta (1994), 'The Interest-based Explanation of International Environmental Policy', *International Organization*, vol. 48 (Winter), 77–106.

UNEP [United Nations Environment Programme] (2000), *Global Environmental Outlook 2000*, United Nations Environment Programme, Nairobi.

UNEP (2002), *GEO–3*, Earthscan, London.

Victor, D. and J. Ausbel (2000), 'Restoring the Forests', *Foreign Affairs*, vol. 79 (November/December), 127–44.

von Moltke, K. (2000), *An International Investment Regime? Issues of Sustainability*, International Institute for Sustainable Development, Winnipeg.

World Bank (2000), *The Little Green Data Book 2000: From the World Development Indicators*, International Bank for Reconstruction and Development, Washington DC.

Part III

The Case Against a World Environment Organization

Chapter 7

Clustering International Environmental Agreements as an Alternative to a World Environment Organization

Konrad von Moltke

1   Introduction

There is widespread consensus that the existing structure of international environmental management needs reform and strengthening. The impetus for this consensus is fourfold: First, the creation of the Commission on Sustainable Development (CSD) at the 1992 United Nations Conference on Environment and Development (UNCED) did not result in the strengthening of international environmental regimes that some may have hoped for. Second, the World Summit on Sustainable Development (WSSD) to mark the tenth anniversary of UNCED created a deadline against which progress could be measured. Third, the continuing need to develop international responses to the challenges of sustainable development has resulted in a structure that is increasingly complex and widely viewed as inadequate to the growing needs that are associated with it. Fourth, the nexus between international economic and environmental policy has grown increasingly powerful, and threatens to result in a deadlock in both trade and environmental negotiations unless some of the organizational issues can be resolved in a satisfactory manner. This growing consensus that international environmental management needs reform and strengthening found its expression in Decision 21/21 of the Governing Council of the United Nations Environment Programme (UNEP).

Yet while this decision launched a process of negotiation within UNEP aiming at the World Summit on Sustainable Development (WSSD) that was held in Johannesburg in August 2002, there remained a remarkable scarcity of realistic proposals on measures that can be adopted. The UNEP process was largely ignored at WSSD. Critics of the proposal to create a World Environment Organization (WEO) still must address the inadequacies of the existing structure for international environment and sustainable development governance. Based on the initial documents from the UNEP process,[1] one of the approaches that are worth exploring is that of 'clustering', that is of grouping a number of international environmental regimes together so as to make them more efficient and effective. This is an issue that has not received systematic attention before now (von Moltke 2002).

Attempts to create a single World Environment Organization face extraordinary difficulties of both a theoretical and a practical nature. From a theoretical perspective 'the environment' is a construct that has no direct institutional equivalent. Since 'the environment' is by definition everything that lies outside human institutions, there can be no single institution that adequately reflects it. The environmental agenda is in fact an agenda of several issues that exhibit distinctively different problem structures. Many of the institutions required to address climate change are different from those needed for hazardous waste management or species conservation, or water pollution control. In addition the need to achieve cooperation between different levels of governance (subsidiarity) is characteristic of environmental management, creating additional incentives to distribute environmental functions. This theoretical observation has been reflected in the extraordinary dispersal of organizations dealing with different aspects of the environment. In fact, most countries have cabinet level environment ministries, but none of these agencies actually is responsible for all aspects of the environment. In practice, the effort to create a World Environment

---

[1]   For the most recent version of documents see http://www.unep.org/IEG/Background.asp.

Organization will be burdened with institutional resistance from existing environmental regimes and by the manifest unwillingness of countries that provide financial support for international environmental measures to give up control over the funds that are made available.

2   Clustering

The current number of international environmental regimes is clearly too large to be optimal. This large number is rooted in the fact that structural differences exist between many environmental problems, thus requiring separate institutional responses. The institutions required to manage biodiversity are obviously different from those needed for hazardous waste, and the institutions for climate change differ in many respects from those for water management, or ocean governance for that matter. Nevertheless it is not possible to argue that the actual number of international environmental agreements—in excess of 300 by some counts—represents an appropriate number from the perspective of effectiveness.

The actual merger of existing international environmental agreements is a daunting task. It has been accomplished but once, when the Oslo and Paris conventions were merged. Yet despite the manifest advantages of a merger and despite the fact that the membership of both agreements was identical and involved a limited number of highly developed states the process of merger took many years to accomplish. The reasons why such a merger does not appear feasible except in singular cases are numerous:

First, the negotiators of the historically latest agreement were presumably aware of the existence of prior agreements with related, or even overlapping, subject matter. Yet they chose to negotiate a new agreement, with new institutions, rather than build on the existing structure. The reasons to do so must have been compelling at the time, and any proposal to change these decisions subsequently must at the very least respond to the reasons that prevailed when negotiations were undertaken.

Second, membership of related or overlapping agreements is rarely identical. Thus key countries party to the Convention on International Trade in Endangered Species (CITES) are not party to the Convention on

Biological Diversity (CBD). Their merger entails the risk of losing parties in one regime without gaining more penetration in others.

Third, even where membership is identical the domestic constituencies supporting related or overlapping regimes may differ. This is most frequently expressed by differences in bureaucratic responsibilities. Thus the agency responsible for the Basel Convention on the International Transport of Hazardous Wastes may not be responsible for the management of toxic substances and thus play a minor role in the Convention on Prior Informed Consent or the Convention on Persistent Organic Pollutants. Unfortunately such differences in attribution can pose problems even within a single agency in a given country.

Fourth, the existence of an international environmental regime frequently gives rise to congruent structures in international civil society—for example scientific groups, commercial interests, or advocacy organizations—resulting in a committed constituency whose very existence may be threatened by proposals to merge, move or abolish a regime. Fifth, in several instances later conventions represent an evolution in thinking about certain environmental problems. Despite addressing related or overlapping problems they may exhibit quite different institutional structures and pursue distinct priorities that a merged regime would have difficulty in balancing. Finally, decisions concerning the location of secretariats are often highly competitive; some countries have shown an active interest in attracting the permanent organization associated with a given regime. Having expended effort to obtain the location of a secretariat in their country, having generally been required to support that secretariat in a variety of ways that required budgetary allocations, the countries concerned have strong stakes of ownership in the secretariat.

In practice any attempt to negotiate all the factors that obstruct a merger, even when it seems logically unimpeachable, will require extraordinary effort while possibly producing modest results in terms of greater effectiveness or efficiency. At the very least it risks the misallocation of one of the scarcest of resources: the negotiation effort of the constituencies involved and the attention of senior policy makers.

Under these circumstances it may be appropriate to seek a variety of institutional and organizational arrangements short of a merger that will increase the efficiency and effectiveness of existing agreements without requiring elaborate changes in legal or administrative arrangements. This is what is meant by 'clustering'. It is important to view clustering as a process and not as a single act, so the immediate task is to create conditions that are conducive to fostering a process of clustering. The assumption is that the experience of working in clusters can give rise to subsequent changes that contribute to further increases in efficiency and effectiveness.

3   The Tools of Clustering

The notion of clustering assumes that there are ways to promote closer integration of related or overlapping international environmental regimes, short of merging organizations. It is worth listing the tools of clustering, even though not all may be applicable to every cluster, and certain clusters may have additional tools that can be utilized.

The Conference of Parties

Most international environmental regimes have a conference of the parties or some similar institution as the ultimate source of decision-making. The conference of the parties meets periodically in locations that are determined from one meeting to the next. Several important options are available with regard to the conference of the parties, precisely because no permanent commitments have been made thus far concerning timing and location.

(a) Co-location. The conference of the parties of clustered agreements can be held simultaneously in a changing location. This would facilitate coordination between the regimes while leaving a range of options open concerning the relationship between these simultaneous meetings, for example consecutive scheduling, joint bureaus, or joint activities relating to civil society.

(b) Permanent location. In addition to deciding to hold conference of the parties simultaneously, it is possible to always hold them in the same location, whether simultaneously or not. This permits the development of an infrastructure to support the conferences of the parties, including the possible creation of specialized missions from member states. One of the

lessons to be derived from the experience of the WTO is the advantage of a single location and the importance of permanent missions devoted to the WTO agenda. These missions have in fact become an integral part of the organizational structure of the WTO, and explain in large measure how the organization manages to cover a wide agenda with a relatively small secretariat. The advantages of holding simultaneous meetings are clear. This would also hold the additional benefit of facilitating developing country participation in the environmental regimes. It would also tend to strengthen the role of member states.

(c) Executive and subsidiary bodies. Many conferences of the parties have executive and subsidiary bodies that meet between sessions of the conference of the parties. The scheduling of these meetings can occur according to a variety of conventions, alternating between a permanent location and a flexible one (as in the case of the World Bank and International Monetary Fund annual meetings), always in alternating locations, or in some rotating pattern with the conference of the parties itself.

     There are numerous permutations that can evolve on the basis of the above variables. While it is theoretically desirable to have conference of the parties meetings occur at the location of the regime secretariat(s), it is certainly not indispensable. Most international environmental regimes currently hold conferences of the parties at locations remote from their secretariat. This practice can be continued. Given that the secretariats of clustered regimes may actually be in several locations, there is no reason to assume that holding the conferences of the parties at the seat of one of them will exhibit particular advantages. It would presumably be possible to establish a service unit common to the clustered regimes at the seat of the conference of the parties to provide essential services on a continuing basis.

Subsidiary Bodies

Most international environmental regimes have a number of subsidiary bodies concerned with scientific and financial matters. It may prove possible to move beyond co-location to a more permanent form of coordination between these bodies. This measure can precede coordination of conferences of the parties or follow it, depending on priorities of the particular cluster. Delay in holding simultaneous meetings or identifying a permanent location for the subsidiary bodies—which can but need not be identical to the location of the conference of the parties—can help to ease the transition and contribute to maintaining the presence of international environmental regimes in a wide range of locations.

Secretariats

All major international environmental regimes have a secretariat to ensure continuity and coordination. These secretariats are often the most visible manifestation of the regime so that efforts at strengthening and coordination tend to focus on them. At the same time, moving a secretariat requires extraordinary effort. The specific role of the secretariats can differ from one regime to another, reflecting both different legal authority and the result of a dynamic development of the regime itself. The organizational arrangements for individual secretariats can also differ widely, even among quite small organizations, depending on whether it is an independent body, located within some larger international organization, revolving between states (like in the case of the Antarctic treaty) or based on a nongovernmental organization. Finally leadership plays a significant role in secretariats, which can acquire certain characteristics as a consequence of the personality of the person responsible for them.

Given all these constraints, the prospects for dramatic reorganization of secretariats appear remote. In practice such reorganization is not as vital as it may appear. Regime secretariats are responsive to a range of factors, including the conference of the parties, domestic and international constituencies, financial arrangements, scientific advice and media pressure, which are more amenable to change than the secretariats themselves. In practice, every cluster is liable to involve several existing regimes with separate secretariats, which will only rarely be in the same location. Consequently solutions need to be found that permit these secretariats to work more closely together, short of actually moving them. Staff exchanges, the use of common staff under certain circumstances, and the aggressive adoption of communications technologies all can serve to alleviate what might otherwise appear as an insuperable problem.

Financial Matters

Purposeful use of financial incentives represents a significant factor in clustering. Like most other measures to promote clustering, the use of financial tools is promising only if it is undertaken consistently by all key parties to an agreement. Nevertheless individual parties may find that it is possible to make appropriate adjustments in their own approach to financial issues relating to regime clusters. While this may not produce the desired changes in the regime as a whole it can increase the efficiency in the allocation of that party's resources and create incentives for other parties to act in a complementary manner.

Most international environmental regimes are supported by voluntary contributions. The power of the purse represents an important tool in situations where a significant group of parties agrees on the need to promote clustering. In these instances the parties that finance the infrastructure of the regime would be justified in using their position to accelerate and give direction to the clustering process.

(a) Regime budgets. The budgets for the operation of individual environmental regimes are generally quite modest—with the exception of the climate regime. Some funds have been generated to support the goals of certain regimes, notably the Global Environment Facility and the Multilateral Fund for the Stratospheric Ozone Regime. These funds are typically not controlled by the regime secretariat and do not contribute to its expenses. Yet taken together the budgets of all regimes in a cluster can be substantial. These include the resources required to ensure the participation of developing countries. All regimes struggle to obtain adequate resources to ensure their operations, with voluntary contributions predominating. Any move to cluster resources for groups of regimes would create powerful incentives for coordination between those responsible for the regimes' finances.

(b) Development assistance. Many international environmental agreements call for the provision of new and additional funds for development assistance. Indeed, UNCED involved an implied bargain that developing countries would participate more actively in international efforts to protect the environment and developed countries would contribute more

vigorously to the funding of relevant activities. The extent to which these commitments have been met has not been tracked but the consensus appears to be that developed country performance in this area leaves much to be desired. Close tracking and active coordination of development assistance funding for certain clusters should generate incentives to ensure the more effective and efficient use of the scarce resources that are available.

(c) Subsidies. Subsidies are an integral part of the environmental policies of any country. Most countries have found that in the early stages of creating essential environmental infrastructure subsidies are necessary to accelerate the process and to drive it beyond the relatively modest parameters that have been set. Such subsidies involve the risks associated with any programme of subsidy—that they become self-defeating, subject to capture by interest groups and ultimately represent an obstacle to the achievement of market-based environmental objectives. Despite these drawbacks, subsidy programmes are an integral part of any environmental strategy, whether open or disguised in a variety of ways. In effect they represent a way to finance environmental conservation that does not have an identifiable market value. The Global Environment Facility is an institution for international subsidies. Its role in a more clustered system needs to be considered carefully. In practice, each cluster involves quite distinct types of activities that require international support. It appears desirable to ensure a closer link between the substantive authority and the project activity than has been accomplished under the current structure.

Electronic Clustering

At least theoretically, modern communications technology offers a range of opportunities for reinforcing the relationship of related and overlapping environmental regimes. In practice, modern technology relies on personal relationships as much as previous technologies so that electronic activities alone entail few substantive benefits. They can, however, provide a powerful tool to support other kinds of clustering activities and facilitate linkages over distance.

Cluster Coordinator

No cluster can function without clear assignment of roles and responsibilities. In many respects this assignment—and the likely conflicts surrounding it—form the heart of any clustering activity. It is critical to ensure that an individual, or a group of individuals, are given responsibility for the work of a cluster. Geographic location is a variable that can be utilized creatively, as can the range of possible organizational affiliations of such individuals or groups. In other words, cluster coordination can occur at the site of one of the secretariats, at the site of joint conferences of the parties, or at a site that offers particular advantages from the perspective of the UN system, New York or Geneva in particular.

In theory, international secretariats are the servants of the member states and the conference of the parties. Yet in practice the need to articulate underlying issues in a continuous manner has given secretariats—and in some instances their respective leadership—roles that transcend this fairly limited notion. Clustering of conferences of the parties will tend to reinforce the role of states in the regimes, in particular if a system of permanent representatives at the location of a conference of the parties emerges. Clusters will, however, need leadership and a visible public presence, particularly where issues of great public saliency are concerned. Striking the right balance in this regard is one of the major challenges of any clustering process.

Implementation Review

International environmental regimes are characterized by a high degree of subsidiarity. In other words, the activities of several levels of governance must work together. From this perspective an active policy of implementation review that encompasses both the national and the subnational levels appears particularly important. One option is to undertake a review of all international environmental obligations of a given country. This creates incentives to strengthen all international environmental regimes, and can also provide important guidance to funding support for implementation in developing countries. An alternate approach would focus on groups of related or overlapping agreements, permitting a more detailed and specific review. In this instance it becomes possible to articulate quite specific performance goals for the period between reviews in relation to a given cluster.

Reviews could proceed along the lines established by the WTO and the Organization for Economic Cooperation and Development (OECD). This involves the preparation of a country report, either by the authorities of the country in question or by the relevant secretariats, or by an agency such as the UN Environment Programme, followed by a country visit by a team of 'reviewers'. The reviewers are chosen in consultation with the country involved and should be given an opportunity to travel as necessary and to meet with any person or groups in the country that they find necessary. The country reports, together with the reviewers' findings, are subsequently discussed in a forum of member states established for this purpose.

Communications

The public image of international regimes is formed to a significant degree by their communications strategy. Clusters can develop a joint communications strategy, including publications and an internet strategy, which can help to strengthen the internal links of the cluster. There is a good deal of duplication of effort between environmental regimes when it comes to communication. In addition, transparency and public participation are critical institutions to promote the effectiveness of international environmental regimes, so efforts to optimize the use of scarce communications budgets must be welcomed.

Capacity Building

Environmental management is institutionally demanding. It requires a large number of effective institutions at the domestic level, and it requires administrative structures that promote cooperation. Since many environmental decisions have potential impacts on a wide range of economic interests there needs to be a highly developed consultative process to minimize such impacts, and there needs to be a review process to ensure that decisions that are taken are appropriate and legitimate. These activities impose significant burdens on domestic institutions in all societies. In developing countries the problems can become insuperable, so that even when the political will exists to promote sustainable development it can prove almost impossible to advance this agenda without significant investments in capacity building. Many international environmental agreements contain provisions concerning special and differential treatment of developing countries and capacity building. Properly conceived, capacity building initiatives can become powerful tools for clustering, conveying the necessary skills and providing a more coherent and effective international environmental management structure to interact with.

4   Creating Clusters

It is common practice to group international environmental agreements by topic, since this is preferable to the only alternative—chronological order—to create some structure in a universe of several hundred agreements. Like any system imposed on a structure that evolved without systematic intent, this requires a certain degree of arbitrary assignment. It is not the purpose of the following grouping to achieve a perfect system to categorize all international environmental agreements. Its intent is to form clusters of agreements not by subject area but by problem structure (von Moltke 1997). While some clusters remain quite predictable, it emerges that some agreements that apparently deal with the same issue do not belong together because of major institutional differences that are rooted in differences in problem definition. Other agreements that appear to deal with institutional issues relevant to most problem clusters—the PIC Agreement for example—in fact addresses only the institutional needs of a single cluster.

The formation of clusters is clearly a matter for broad discussion, careful consideration and full negotiation. It is not the kind of issue that is amenable to analytical approaches alone since only the process of negotiation can ensure that all important stakeholders are heard and all significant issues are given due consideration.

The Conservation Cluster

The conservation cluster[2] is characterized by two major global conventions whose relationship remains a matter of discussion, and a number of other global and regional agreements that are at present poorly integrated. Three of the conventions mark the evolution of international approaches to conservation. The Ramsar Convention is largely devoid of substantive international obligations and sees its primary focus at the national level. CITES addresses the most obviously international dimension of conservation—trade in endangered species. At the same time it has become the focus of an extraordinary scientific effort to identify and assess potentially endangered species of all kinds. The Convention on Biological Diversity seeks to achieve a fully integrated approach to conservation, recognizing both human use and the need to protect entire ecosystems, including both in situ and ex situ conservation techniques.

While the complex would clearly benefit from a significant organizational overhaul, each regime has developed its own constituency, which is frequently willing to defend its independence. Integration requires a comprehensive understanding of the issues and of the role each of the regimes can play in developing an international response to the imperative of conservation.

To represent a significant step forward, a Global Conservation Regime would need to provide additional institutional support to the protection of wetlands and other critical habitat and incorporate most regional conservation activities, several of which deal with migratory species that are not covered by the global agreements.[3] The lack of integration between the global and regional conservation regimes, which do

---

[2] World Heritage Convention; Convention on Biological Diversity (CBD); Bonn Convention on the Conservation of Migratory Species (CMS); Convention on International Trade in Endangered Species (CITES); Ramsar Convention on Wetlands. The UN Convention to Combat Desertification (CCD), the FAO International Undertaking on Plant Genetic Resources, and the International Tropical Timber Agreement.

[3] The Bonn Convention on the Conservation of Migratory Species has not evolved into the universal framework that its drafters envisaged, lacking some key members and without a strong civil society constituency.

not even operate according to a common understanding of the issues and an accepted distribution of roles, is one of the major current challenges facing international conservation efforts. An initial step could be the identification of critical conservation areas that are of importance to all or most of the conservation regimes and to focus resources on these areas. This is itself a matter for international negotiation rather than expert analysis.

The Global Atmosphere Cluster

The two agreements in this cluster[4] involve complex institutional arrangements. Indeed, one of the burdens on the climate regime is the tendency of some observers to assume that the ozone regime represents a template on which to build. In practice the ozone regime is based on a relatively traditional agreement that identifies pollutants and then takes steps to reduce their production, use and emission to levels that are deemed acceptable. Since this involves a class of industrial chemicals that are used in the production of a range of goods, the ozone regime demands a good deal of adjustment from manufacturers but has little direct impact on the end users of the affected products, except perhaps with regard to price. The climate regime deals with several 'pollutants' that are ubiquitous, indeed that are an integral part of life. Control of these substances requires structural change at all levels of society and economy. The resulting regime is essentially an investment regime that seeks to reduce emissions by shifting the focus of public, corporate and private investment.

Despite these differences, the two global atmospheric regimes represent an obvious cluster. Yet the prospects for achieving significant progress are burdened by the historical decision to set up the United Nations Framework Convention on Climate Change (UNFCCC) as an essentially independent organization within the UN system rather than assign it to one of the competing claimants—primarily UNEP and WMO.

---

4   UNFCCC and Kyoto Protocol; Vienna Convention and Montreal Protocol on Substances that Deplete the Ozone Layer. The international agreement on long-range transboundary air pollution (LRTAP) exhibits a significantly different problem structure.

The UNFCCC is already one of the largest convention secretariats in the United Nations, and the complexity of the issues it faces suggest it will grow further in importance.

The Hazardous Substances Cluster

All of the agreements in this cluster[5] are managed by UNEP, so that it already exhibits certain coherence. The control of hazardous substances is essentially the control of the products of a few industries, primarily chemicals and minerals production. A large portion of these industries is located in or controlled from OECD countries. Consequently ways must be found to better integrate the OECD work in this area into a broader global framework.

The recently concluded Convention on Prior Informed Consent (PIC) and the Convention on Persistent Organic Pollutants (POP) represent essential building blocks of this cluster. With these in place it should be possible to move towards greater integration, but for the obstacles outlined above. In many countries the agencies responsible for hazardous wastes are not identical to those responsible for the control of toxic substances. Frequently waste management is the responsibility of states or provinces while toxic substances control is invariably the responsibility of national authorities.

The World Health Organization (WHO) has done important work on heavy metals in the environment that reflects the priorities of the health professions. The activities of the Food and Agriculture Organization of the

---

5   Bamako Convention on the Ban of the Import into Africa and the Control of Transboundary Movement and Management of Hazardous Wastes within Africa; Basel Convention on the Control of Transboundary Movements of Hazardous Wastes and their Disposal; Convention on Civil Liability for Damage Caused During Carriage of Dangerous Goods by Road, Rail, and Inland Navigation Vessels; Rotterdam Convention on Prior Informed Consent (PIC); Convention on Transboundary Effects of Industrial Accidents; Waigani Convention to Ban the Importation into Forum Island Countries of Hazardous and Radioactive Wastes and to Control the Transboundary Movement and Management of Hazardous Wastes within the South Pacific Region; Stockholm Convention on Persistent Organic Pollutants (POPs). The FAO Code of Conduct on the Distribution and Use of Pesticides could be included since it has a similar problem structure. Its institutional approach is, however, hardly comparable.

United Nations (FAO) concerning pesticides and the work of the Codex Alimentarius Commission on residues in food (a joint undertaking of FAO and the WHO) are relevant but will presumably remain outside the core cluster. All of these activities would need to be reflected in the hazardous substances complex in some fashion.

An additional challenge in the hazardous substances complex is presented by the need to integrate the OECD chemicals process, which is essentially a regional international agreement. This is an area in which the use of creative institutional arrangements is required to ensure the integrity of the OECD process, which has been won with great difficulty, while better integrating it into the wider global structures.

The Marine Environment Cluster

There are a large number of agreements[6] that deal with the marine environment involving several organizations, including the International Maritime Organization (IMO), UNEP, and the UN Convention on the Law of the Sea (UNCLOS). The IMO manages agreements concerning pollution from ships; UNEP manages the regional seas programme; and the UNCLOS Secretariat handles the broader legal framework. The approach of each group of agreements is markedly different.

The UNCLOS is the most classic of all fields of international law, carrying the encrustation of several centuries. While it represents the framework within which all other marine activities are undertaken it has a mixed record of effectiveness with regard to matters that concern the environment. It has, however, given rise to the Law of the Sea Tribunal, a unique institution in that it parallels the work of the WTO dispute settlement process but with a higher degree of predictability and transparency.

Over a period of several decades, the IMO has succeeded in bringing the problem of intentional discharges of oil from ships into a

---

6   IMO Conventions; UNEP Regional Seas Conventions; OSPAR Convention for the Protection of the Marine Environment of the North-East Atlantic; Helsinki Convention on the Protection of the Baltic Sea.

management structure that holds out the prospect of being effective. It has reduced the pollution risks associated with marine accidents by steadily improving the design of the ships carrying the most hazardous cargoes. It has established rules concerning the intentional discharge of oil from ships, in particular for deballastage, which can address what is the largest source of oil pollution from ships, even though enforcement can be difficult. The IMO has always struggled with the problems posed by flag state jurisdiction, and some of its advances are due to innovations limiting the reach of this principle, for example by permitting the introduction of port state jurisdiction over certain activities.

UNEP's regional seas programme addresses the broader environmental agenda, including the dumping of waste at sea—an activity that has largely been stopped—and the exceedingly difficult challenge of controlling land based pollution so as to protect the marine environment. In principle, the regional seas programme also addresses issues of coastal zone management, an area that is particularly burdened in most countries by the existence of numerous competing jurisdictions. The UNEP programme is hampered by its technical complexity and the fact that it imposes demanding requirements on national governments that are not always willing or able to live up to them.

The current effectiveness of the agreements in this complex is mixed. Further strengthening of port state jurisdiction and of the rights of states to control their exclusive economic zones may prove helpful. The creation of an effective cluster in this area would require a very substantial amount of negotiating effort.

The Extractive Resources Cluster

This is the most difficult of all environmental issues, and the one with the largest potential impact on the trade regime.[7] At present, international commodity regimes are largely mixed public/private structures designed to

---

[7]   This complex includes most forestry agreements and public/private initiatives such as the Forest Stewardship Council or the Marine Stewardship Council. It also encompasses fisheries and agreements concerned with the environmental impacts of agriculture.

extract natural resources and to distribute them globally, for example the banana regime, the aluminium regime, the cotton regime, or the forest products regimes. Attempts to introduce environmental criteria, let alone sustainable development criteria, into these regimes have met with limited success. Yet all of these regimes have a significant sustainable development dimension. The environmental impacts are largely focused at the extractive end, while funding for each regime, including for sustainable development, needs to come from the consumer rather than from public sources. Consequently the problems of these regimes relate as much to the functioning of international markets as to the possibility for developing international agreements covering their sustainability.

International markets for goods—in particular for commodities—have seen a dramatic transformation through the multiple forces of liberalization, including both multilateral negotiations in the GATT/WTO, domestic measures under pressure from the Bretton Woods institutions, or unilateral measures adopted by individual countries. As a result the role of governments in these markets has been significantly reduced. Consequently governance of these markets now involves private actors to an even greater extent than before. This creates notable opportunities for innovative structures, for example the development of partnerships that include all stakeholders, public and private, producers, traders, and consumers, so as to create agreements that address all important stages of a product chain.

Regional Clustering

A significant number of environmental issues are based on the use and management of land. While many of these issues have an international dimension it is typically not global in character, affecting neighbouring states within a regional land pattern. River basins are an obvious example of such linkages. In addition, long range air pollution, while theoretically a global phenomenon, in practice requires regional responses.

Addressing these issues requires a continuous balancing of conflicting policy priorities, involves high levels of inter-jurisdictional cooperation, and is often viewed as particularly sensitive to concerns of security and sovereignty. All of these factors render a global regime highly impractical, yet it is necessary to ensure a basic level of international cooperation. The response needs to be some form of regional clustering.

The basic structures for such clustering exist in Europe, based on the UN Economic Commission for Europe and a suite of agreements that have evolved steadily following conclusion of the Helsinki Accords in 1977 and the civic revolution that swept Eastern Europe ten years later.[8] Conditions are much less well developed in other regions and in some, such as Asia and the Pacific, even first steps appear not to have been taken.

5   Joint Institutions

Several institutions[9] recur throughout the structure of international environmental management. International environmental regimes are characterized by a large variety of institutions. The reasons are to be found in the structure of environmental problems that require social and economic institutions to address a phenomenon that is governed by the laws of nature. As a consequence international environmental regimes have exhibited a remarkable degree of innovation as they have struggled to match their institutional arsenal to the structure of the problem they attempt to address.

Some institutions, in particular those that translate science into policy and that seek to assess environmental conditions in a systematic manner, are pervasive throughout international environmental regimes. Even when not every regime utilizes a particular institution, it is worth considering the options for creating crosscutting rules to ensure consistent application and to develop new organizational structures to promote greater efficiency and effectiveness. This appears as an area of activity for a broadly based organization, such as UNEP.

---

[8] See also Kirton, this volume, on economic and environmental cooperation in the North American region.

[9] The term 'institutions' is used here in its strict technical sense to denote the rules of the game that characterize a regime. Thus 'property' is an institution but UNEP is an organization.

Science Assessment

Science assessment is the interpretation of research for policy purposes. Most countries use science assessment institutions to mediate the complex relationship between scientific research and public policy. Arguably the most characteristic institution of all environmental regimes—because without scientific research there can be no environmental management—science assessment offers a range of options for the clustering process at a universal level.

Few international environmental regimes have the necessary resources to undertake science assessments of their own or even to review science assessments undertaken at national level with a view to identifying the specifically international interest. Apart from the Intergovernmental Panel on Climate Change (IPCC), there are no fully developed science assessment mechanisms at the international level. The resources required to undertake full-scale science assessment on a major issue of international environmental concern are very significant. It makes much more sense to focus the necessary resources on one or two regimes at any one time rather than distributing them widely, as now occurs. Consequently a structure needs to be devised that can draw on the best scientists worldwide in changing fields of research. The model would be the US National Research Council (a branch of the National Academy of Science), which is required by charter to provide government with advisory services (for pay) yet manages to maintain its independence and its ability to identify appropriate participants in its panels from a range of disciplines.

Monitoring and Environmental Assessment

Specific environmental measures are based on numerous assumptions about environmental conditions, the need to adopt measures, and the impact of these measures on environmental conditions. These assumptions are fraught with many uncertainties, attributable in particular to lack of scientific knowledge or lack of information about actual environmental conditions. Responsible policy-making will ensure that these assumptions are tested on a continuous basis, primarily through further research and through an appropriate programme of monitoring and environmental assessment.

Monitoring and environmental assessment are also required for international environmental policy. In practice, much of the monitoring will be undertaken at national or subnational levels, but it is important to ensure comparability of data and coordination of monitoring schedules to ensure that international concerns can also be addressed. Some countries may require assistance in setting up and funding monitoring systems. The actual assessment process needs to have an independent international component.

Monitoring and assessment are cross-cutting activities. It does not make sense to engage in separate monitoring for each cluster since many of the pollutants of concern—in particular heavy metals and persistent organic pollutants—migrate from one environmental medium to the next and must be monitored on an integrated environmental basis. Consequently this represents an institution that is best entrusted to a universal organization. The current system of monitoring and assessment needs to be significantly strengthened. This requires both additional funding and a process to set priorities and to eliminate duplication of effort.

Transparency and Participation

Transparency and participation have emerged as central institutions for all environmental regimes, a reflection of both scientific uncertainty and subsidiarity: public authorities, even local authorities, cannot have detailed knowledge about environmental conditions in specific locations, and some environmental phenomena emerge in the field before they become apparent in the laboratory. The institutions of transparency and participation have become the standard response to this dilemma. Indeed, most concerned with environmental issues have come to expect certain levels of information and access as an integral part of all environmental regimes.

Given the importance of these institutions in environmental affairs it is remarkable that few formal rules have been adopted in international environmental agreements to formalize them. More recent agreements tend to include the necessary provisions. Many environmental regimes tend to rely on established practice and informal understandings. The Århus

Convention on Access to Information, Public Participation in Decision Making and Access to Justice in Environmental Matters (1998) represents a first step towards developing universally applicable rules—although they do not apply to international agreements but rather are binding on countries that are party to the Convention only.[10]

An attempt to develop a broader international agreement applicable to all international environmental regimes is necessarily fraught with risk: asked to codify current practice some countries are likely to seek to limit it.

Implementation Review

In most international environmental agreements implementation review is the responsibility of the conference of the parties. One instrument to promote greater coherence among these regimes, and within their member states in matters of international environmental management, is to institute joint implementation review of individual countries. Such a review process would require some level of cooperation between the regimes involved and at the same time foster greater coherence in the implementation efforts of the countries that are being reviewed.

This is an area where the example of the GATT/WTO may be helpful. The Trade Policy Review Mechanism has evolved patterns of work that appear to be acceptable to member states while still generating information that can be useful to other states and at the international level. The essential characteristic of an Environmental Policy Review Mechanism would, however, be its ability to address all aspects of international environmental management rather than basing its approach on a limited number of agreements alone. In particular it would need to be able to include regional agreements or the instruments of the OECD.

---

10   The Århus Convention was adopted in the context of UN-ECE, and has not been signed by all members of that body.

Dispute Settlement

Dispute settlement (based on legally binding rules) is the issue most frequently mentioned as distinguishing trade regimes from environmental ones. It is also frequently mentioned as an area where environmental regimes could benefit from further institutional strengthening. Yet there is no evidence from environmental regimes themselves that this is an area of great current concern. In practice the International Court of Justice (ICJ) serves as a dispute settlement mechanism of last resort. Not only has it not been used, there are some cases where it has been explicitly avoided and in those instances alternative forms of dispute settlement have emerged.

However, the assumption that stronger dispute settlement in environmental agreements will relieve pressure from the trade dispute settlement process assumes a parallelism between trade and environment that does not exist. In the trade regime, dispute settlement is the premier implementation tool—and to a significant degree the pathway by which interpretation of the agreements can be adjusted (Jackson 2000)—and consequently the place to which issues such as the environment must migrate. Environmental regimes pursue effectiveness and implementation through entirely different institutions (transparency and participation in particular) and there is no reason to assume that the availability of a reinforced dispute settlement mechanism will change that in any way.

The nature of legal obligations entailed in multilateral environmental agreements—and the structure of the ensuing regime—is such that environmental regimes rarely generate the kind of dispute that is characteristic of the WTO system. Appropriate remedies would be difficult or impossible to craft. When such disputes arise, they tend to migrate directly to the conference of the parties of the relevant agreement since they require a process of negotiation rather than adjudication. It is certainly possible to interpret the long and arduous process on listing, re-listing, and possibly delisting the African elephant in CITES, which several times worked its way through the institutions of the regime to the conference of the parties and back, as a process of dispute settlement.

Environmental disputes between private parties represent a challenge to the international legal system. The protracted dispute about salt pollution of the Rhine is emblematic for these issues. The issues such private disputes raise are issues of general international law rather than of the institutions of environmental regime.

The disputes that can arise in international environmental regimes concern lack of implementation of domestic environmental law, whether or not it implements international obligations. One state can hardly launch a complaint about such non-compliance against another. No state is flawless in this regard. The adequacy of domestic implementation is a matter that requires careful assessment. It is not a matter of interpreting international legal obligations and the remedy is not a change in the rules, domestic or international, but a change in the functioning of domestic institutions.

The only institution that has been identified to launch such disputes is that of citizen complaints. This institution has been used in the European Union and in NAFTA with mixed results but it has certainly strengthened international environmental management. This practice, carefully defined, together with forms of alternative dispute settlement such as mediation procedures could strengthen international environmental regimes while reflecting their particular structure and needs.

6   National Coordination

For many years observers have decried the lack of national coordination of positions in different international regimes. Certainly an increase in national coordination holds the promise of promoting clustering. Yet the obstacles are significant and are not accessible to international negotiations. The one international instrument that may be able to promote national coordination is an integrated process of implementation review (see above). There are essentially three obstacles to greater national coordination: domestic distribution of responsibilities, development of constituencies, and the politics of coordination. Greater national coordination can only be expected if all three factors are addressed at the same time.

Domestic Distribution of Responsibilities

The greatest obstacle to coordination is the domestic distribution of environmental responsibilities. One of the paradoxes of the debate about a

WEO is that it occurs despite the fact that no country has established a domestic agency that covers all the issues that would be addressed by a WEO. The reasons for this state of affairs are manifold. 'Environmental management' in practice involves a significant number of policy areas that share a concern for impacting the environment through changing human behaviour but which exhibit a widely differing problem structure. The protection of biodiversity and the management of hazardous wastes are both considered part of the environmental agenda, yet they require entirely different policy strategies. Similarly, the protection of the marine environment and the reduction of air pollution are closely linked—because atmospheric deposition is a principal source of marine pollution—yet they entail quite different management structures. It is consequently quite reasonable to assign responsibility for biodiversity to one agency and for waste management to another. Indeed, even when both are undertaken from the same agency they may in practice have little routine overlap, except in agency leadership.

In addition to exhibiting a different problem structure at the national level, environmental issues are subject to different levels of subsidiarity. Some issues such as land use are deeply rooted in local governance. Other issues, such as the management of watershed, exhibit regional structures. Yet other issues, such as the control of hazardous chemicals, are typically of national concern. Finally some issues, such as atmospheric pollution, can be addressed in a variety of ways depending on the degree of centralization or decentralization that is typically preferred by a country. With such a variety of possible approaches, it is hardly surprising that every country has an essentially unique pattern of responsibilities.

The environmental agenda grew incrementally, sometimes over a period of decades. In most developed countries the roots of water pollution control and the management of industrial facilities reaches back into the 19th century. Biodiversity protection on the other hand is an issue of the last decade of the twentieth century. The notion that 'the environment' as a whole requires integrated management did not emerge until the 1970s. Countries responded differently to these changing perceptions. While most countries, with the exceptions of the United States and Russia, have cabinet level environment ministries, none has one that encompasses all aspects of the environment as it is now understood.

The traditional approach to a need for coordination of national positions in international fora is to assign responsibility to the foreign affairs agency. This is possible where the issues concerned do not involve changes in domestic legislation and the responsibilities of subnational units in a federal system. In those instances, foreign affairs agencies have few of the needed skills to balance international needs against domestic regulations and priorities. In many countries this has led to wholesale delegation of international responsibilities to the various environmental agencies. Coordination may be better in countries where that has not occurred but at the price of poor integration with domestic policies.

Constituencies

The adoption of an international environmental agreement almost always engenders the emergence of a complex regime that includes many actors beyond the states party to the agreement. Several groups from civil society are typically involved, including scientists, industry and commerce and advocacy groups of all kinds. Even government agencies other than those primarily responsible for an issue can find themselves involved indirectly. This phenomenon is one of the most important sources of effectiveness of international environmental agreements, since it permits the regime to establish deep roots in various countries. At the same time, the existence of these constituencies can become a significant obstacle once there is a call for 'coordination. Moreover, these constituencies are not readily influenced by the international negotiation and are frequently in a position to create roadblocks to the process.

In some instances, there are also phenomena of bureaucratic clientelism, in the sense that each bureaucracy has a commitment to 'its' international regime, which it views as a vehicle to advance its own agenda, both internationally and domestically. Frequently it is the international dimension that enables the agency in question to attract policy attention from the highest levels of government, and the prestige and resources that can flow from that.

Politics of Coordination

Domestic coordination carries a price. A government that engages in a domestic process of coordination must make hard decisions, at least in the sense of decisions that may displease some constituency or another. Such decisions carry an immediate political price since it involves a clear declaration of government policy in one form or another. Once the government in question reaches the international level, with its carefully coordinated position, it finds that it is but one voice among many. Only very few international actors are able to impose the domestically established compromise on the international process. This has been true even of the United States when it comes to environmental policy issues. Moreover such actors are the most unwelcome of negotiating partners since they are liable to present their domestic positions on a 'take it or leave it' basis, being unwilling or unable to engage in real negotiation. In other words, governments that have carefully coordinated positions are less likely to engage in productive negotiations.

7   How to Begin

The first—and the last—step are the hardest parts of any policy process. The risks are greatest when the first step is taken; and the negotiation process will typically leave the most difficult decisions to last. For this reason every international negotiation—and clustering is unquestionably first and foremost a negotiation process—needs 'champions', countries that have an interest in promoting a certain outcome and are willing to invest some political capital in achieving it. Only the existence of such champions enables international negotiations to lead to outcomes that represent not simply the lowest common denominator of the countries involved. Clusters will also need champions.

The burdens of being champion are such that most processes require no more than a single champion. When more than one appears, this is mostly due to domestic consideration, that is, more than one government feels a need to appear as a champion of an issue at the international level, than of the negotiation itself. Within most negotiations, countries are willing to ally themselves with a champion once it has been identified. This reduces the burden of leadership. Traditionally the country where a secretariat is located has been viewed as the natural champion of a given regime, with the exception of Geneva and New York, which are seats of the United Nations and viewed as relatively neutral in character. One of the problems that UNEP faces, is that Kenya is not an effective champion of its interests in the international system. This is one of the reasons for the current dispersion of secretariats outside UNEP's Nairobi headquarters.

The essential first step in clustering is consequently the identification of champions for various clusters. The existence of several potential clusters suggests that several opportunities exist for championing a cluster. Without such champions, none of the clusters is likely to become reality.

8   Conclusions

The concept of clustering as presented here is not a grand design. There is no need to undertake a specific number of steps to launch the process. It can occur in only one dimension or it can become a more systematic phenomenon. If there have been few moves towards undertaking clustering thus far that is presumably related to the existence of other more pressing priorities—or to a sense that the current structure of international environmental governance is working better than its critics tend to assume.

Existing institutional arrangements are famously resistant to change, in particular at the international level. While the rule of consensus makes it difficult to establish new regimes, the rules of consensus also makes it difficult to change them once they have been established. Consequently most existing organizations tend to prioritize activities that will promote their continued evolutionary development. This is one of the principal reasons why the debate about a WEO has not advanced very far. Clustering is presumably also viewed as an activity that goes beyond this limited ambition.

The hypothesis that the existing structure is not as dysfunctional as many observers assume is a surprising one. The current situation certainly offers a number of advantages. There is some advantage in being viewed as insignificant. Many of the institutional innovations undertaken by international environmental regimes would not have been allowed to go forward if governments had taken them more seriously. Now that they exist, those involved in managing the different organizations and regimes presumably seek to avoid putting the hard-won gains at risk, even if they appear modest. Moreover the large number of small environmental regimes has actually mobilized large overall resources. If all of these were put together, it is likely that they would represent the single largest international regime, in particular when one takes into account the very large number of actors in civil society with a strong commitment to their functioning. Ultimately this construction of environmental governance by subterfuge will have to move more into the mainstream of international society if it is to achieve ambitious environmental goals.

If clustering occurs it probably will not be called that, since the concept is of an analytical construct rather than a political and administrative reality. The need to respond to the trade and environment agenda that has been articulated in the World Trade Organization (WTO) has proven an impetus to cooperation between the affected secretariats on this particular issue. To a certain degree, the Millennium Ecosystem Assessment can be viewed as a clustering phenomenon for the conservation regimes. Thus clustering may occur and only be recognized as such at some subsequent stage.

References

Jackson, J. (2000), 'The Legal Meaning of a GATT Dispute Settlement Report: Some Reflections', in J. Jackson, *The Jurisprudence of GATT and WTO. Insights on Treaty Law and Economic Relations*, Cambridge University Press, Cambridge (UK), 118–32.

Kirton, J. (2005), 'Generating Effective Global Environmental Governance: The North's Need for a WEO', in F. Biermann and S. Bauer (eds), *A World Environment Organization. Solution or Threat for Effective International Environmental Governance?*, Ashgate, Aldershot, 145–172.

von Moltke, K. (1997), 'Institutional Interactions: The Structure of Regimes for Trade and Environment', in O. Young (ed.), *Global Governance. Drawing Insights from the Environmental Experience*, MIT Press, Cambridge (Mass.), 247–72.

von Moltke, K. (2001), 'The Organization of the Impossible', *Global Environmental Politics*, vol. 1 (1), 37–42.

von Moltke, K. (2002), *Whither MEA's? The Role of International Environmental Management in the Trade and Environment Agenda.* Report for Environment Canada [available at www.iisd1.iisd.pubs.html].

Chapter 8

Reforming International Environmental Governance: An Institutional Perspective on Proposals for a World Environment Organization

Sebastian Oberthür and Thomas Gehring*

1   Introduction

A world environment organization (WEO) has been proposed by analysts and policy-makers alike to remedy existing problems of international environmental governance. Despite significant progress in the past decades, sustainable development has not been realized. International environmental problems such as the loss of biological diversity, climate change or the dispersion of persistent hazardous chemicals remain largely unresolved (UNEP 2002). The creation of a full-fledged international organization is expected to strengthen international environmental governance, as did the establishment of the World Trade Organization

*   A similar version of this chapter appears as Sebastian Oberthür and Thomas Gehring, 'Reforming International Environmental Governance: An Institutionalist Critique of the Proposal for a World Environment Organization', *International Environmental Agreements: Politics, Law and Economics*, vol. 4 (4) (2004), in press. It is reprinted here with kind permission of Springer Science and Business Media.

(WTO) for the liberalization of international trade (Runge 2001; Charnovitz 2002). A WEO could provide a common roof for a number of the existing multilateral environmental agreements and form a new 'gravity centre of international environmental policy-making'.[1]

Today, international environmental governance takes place predominantly within numerous independent institutional arrangements. A study conducted for the UN Conference on Environment and Development in Rio de Janeiro in 1992 identified more than 125 separate international environmental regimes (Sand 1992; also Charnovitz 1996). Five additional environmental agreements have been concluded on average per year thereafter (Beisheim et al. 1999, 350-51). International agreements regulate virtually all important regional or global environmental problems. While they have in many cases resulted in remarkable progress, difficulties still abound (e.g. Haas et al. 1993; Young 1999; Miles et al. 2001).

Proposals for the establishment of a WEO form part of a broader policy discussion on reforming the institutional framework of international environmental governance (see Charnovitz in this volume). The Global Ministerial Environment Forum (GMEF) of the UN Environment Programme (UNEP) aims at improving coordination between international treaties and other institutions relevant to the environment. Options reviewed by the GMEF include proposals to cluster multilateral environmental agreements, i.e. to integrate several agreements or certain of their parts (Oberthür 2002; von Moltke in this volume). The World Summit on Sustainable Development convened in Johannesburg in summer 2002 endorsed the efforts of the GMEF.[2]

The idea of a WEO is rooted in dissatisfaction with the current arrangements of international environmental governance and, more importantly, with the lack of effective environmental protection it has achieved so far. Proponents expect a WEO to help overcome in particular

---

[1]   Esty (1994); Biermann (2000, and this volume); German Advisory Council on Global Change (2001); Esty and Ivanova (2001; 2002); Whalley and Zissimos (2001); for an overview and further references see Biermann (2002, 298), Charnovitz (2002, 324-29), and Bauer and Biermann in this volume.
[2]   Paragraph 140(d) of the Johannesburg Plan of Implementation (available at http://-www.un.org/esa/sustdev/documents/WSSD_POI_PD/English/POIToc.htm).

three major problems of international environmental politics (e.g. Biermann 2000; the contributions by Charnovitz, Biermann and Kirton in this volume). The cumbersome process of setting binding international standards for the protection of the environment might be facilitated by bargains across issue-areas and policy fields (Whalley and Zissimos 2001 and 2002). Supervision and enforcement of the implementation of international environmental commitments might be enhanced, partly through the mobilization of additional resources for transfer from North to South to assist developing countries (e.g. Biermann 2000; Esty and Ivanova 2002). In addition, disruption of international environmental governance caused by non-environmental institutions such as the WTO (Runge 2001; Brack 2002; Charnovitz 2002) and by tensions between different international environmental regimes might be mitigated or avoided. The Kyoto Protocol's potential for providing incentives for forestry activities that maximize carbon sequestration while compromising the objectives of the Convention on Biological Diversity (Pontecorvo 1999; Jacquemont and Caparrós 2002) provides an example of tensions between environmental regimes. Finally, a WEO is expected to collect and distribute environmental data and analyses (e.g. Esty and Ivanova 2002; Biermann 2000).

So far, the discussion on the merits of a WEO lacks conceptual foundation. Advocates of a WEO regularly fail to demonstrate why we should expect a WEO to fulfil the aforementioned functions more effectively than the existing institutional arrangements or potential other alternatives (Najam 2003 and this volume). This would require a conceptual foundation enabling us to assess the governance capacity of varying international institutions. This chapter is intended to provide conceptual foundation for the debate by examining the principal contribution that a World Environment Organization could make to enhance international environmental governance from an institutionalist perspective. In particular, we employ cooperation and regime theory, which have greatly advanced our understanding of governance through international environmental institutions over the past 20 years or so (e.g. Haas et al. 1993; Gehring 1994; Young 1994; Victor et al. 1998). We do not engage in a comprehensive review of individual proposals for establishing a WEO. Instead, we investigate, on the basis of cooperation and regime theory, ways in which a WEO might modify international environmental governance.

We make three claims. First, we argue that the establishment of an international *organization* alone in a policy field currently populated by *regimes* cannot be expected to significantly improve environmental governance because there is no qualitative difference between these two forms of governance institutions. Organizations do not have at their disposal additional properties or instruments that are relevant for successful governance and that would not be available in international regimes (Section 2).

Second, we submit that any significant improvement of international environmental governance through institutional re-arrangement can rely on a modification of decision-making procedures and/or a change of institutional boundaries. Modification of decision-making procedures changes the ability of actors to influence outcomes. Modification of the boundaries of the issue-area governed affects opportunities to link issues and conclude mutually beneficial deals. In contrast, an institutional reform cannot be expected to help ensure an effective enforcement of international environmental commitments. Irrespective of decision-making procedures and institutional boundaries, non-cooperating states cannot be excluded from the benefits of international environmental governance, because it is predominantly about the protection of global or regional commons (Section 3).

Third, based upon the analysis of three principal models we argue that a WEO cannot be at the same time realistic, significant and beneficial for international environmental governance. A WEO formally providing an umbrella for existing regimes without modifying existing decision-making procedures would be largely irrelevant. A WEO integrating decision-making processes of existing regimes so as to form comprehensive 'world environment rounds' of intergovernmental bargaining would be largely dysfunctional and prone to a host of negative side-effects. A 'supranational' WEO including large-scale use of majority decision-making and far-reaching enforcement mechanisms across a range of environmental issues might considerably enhance international environmental governance, but it appears to be grossly utopian. From an institutionalist perspective, the creation of a WEO may therefore result in efficiency gains at best (common use of secretariats, coordination of reporting). Political and financial resources available promise greater return if invested in advancing decision-making in existing environmental regimes (Section 4).

2   The Distinction between Regimes and Organizations: Fiction Rather than Reality

The establishment of a World Environment Organization cannot per se be expected to improve international environmental governance. Relevant proposals are in large part inspired by the fact that international environmental politics is predominantly based upon hundreds of separately established international regimes and does not possess a central organization as do other areas such as international trade (Runge 2001; Esty 1994). However, there is no evidence supporting the claim implicit in these proposals that an organization might per se be more powerful or better suited for successful governance than a regime (Najam 2003 and this volume).

International environmental governance is already supported by international organizations and international regimes. Although the latter prevail, both forms are relevant. An international regime is generally based upon one or several international treaties and related instruments. Hence, international climate policy is made within the global climate regime based upon the UN Framework Convention on Climate Change and its Kyoto Protocol. Other regimes address such problems as the depletion of the ozone layer, transboundary trade in hazardous wastes, trade in endangered species (CITES), or the protection of regional seas such the North East Atlantic, the Baltic or the Mediterranean seas. International organizations are commonly considered as actors (Abott and Snidal 1998) and defined by reference to their secretariats and their ability to enter into legal contracts (e.g. Young 1994, 163-83; Keohane 1989, 3-4; Barnett and Finnemore 1999). They also play an important role in international environmental governance. The International Maritime Organization (IMO) hosts a number of agreements for the protection of the marine environment, WHO elaborates air quality standards, FAO manages fish resources. Last but not least, UNEP—despite the fact that it is not a specialized agency but only a UN programme—has promoted the elaboration of several international environmental regimes since the early 1970s.

International regimes and international organizations are in many respects very similar (Simmons and Martin 2002, 194) and do not significantly differ in their governance capacity. They both constitute 'persistent and connected sets of rules and practices that prescribe behavioural roles, constrain activity, and shape expectations' (Levy, Young and Zürn 1995) and they both provide their members with the necessary fora and communication channels to elaborate collectively binding decisions in accordance with established procedures (see also Abbott and Snidal 1998, 15-16). They also do not differ systematically with respect to their patterns of governance. While decision-making in many international organizations remains limited by the requirements of consensus, the procedures in various international environmental regimes are surprisingly far-reaching and include delegation of decision-making authority (see below).

The formal characteristics of the international organizations mentioned above are not suited for a clear-cut distinction and they do not refer to elements that matter for effective governance. While the ability to enter into an external contract under international law may be relevant for some specific tasks, it does not generally increase the governance ability of an organization, because successful governance regularly depends on establishing cooperation between the members of an institution and thus on internal decision-making. Likewise, the success of international governance cannot be attributed to the existence of secretariats. For example, the particular strength of the WTO is usually not attributed to its secretariat but to its decision-making apparatus, in particular to its rigid and influential dispute settlement mechanism (Hoekman and Kostecki 1995, 44-50; Jackson 1999, 107-37).

What is more, if one were to equal international organizations with secretariats, the distinction between international regimes and organizations would become completely blurred. Not only organizations, but also regimes comprise their own bureaucracies. Consequently, both regimes and organizations comprise substantive rules and obligations as well as some sort of a secretariat. The creation of a WEO would amount to the integration of the secretariat services of different existing institutions, for example UNEP and a number of international environmental agreements. Such a step could hardly be expected to significantly contribute to mitigating any of the major deficiencies of international environmental governance.

International organizations do not generally govern broader issue-areas than international regimes, and they are not more successful. Many established international organizations, such as WTO, the Food and Agriculture Organization of the UN (FAO), WHO or the ILO govern large issue-areas. However, others are more specialized, such as the World Customs Organization (WCO) that administers customs codes used in international trade. In contrast, some international regimes such as the regime on climate change and the 1982 United Nations Convention on the Law of the Sea (UNCLOS) manage an enormous scope of inter-connected issues without establishing a formal organization. At the same time, effective organizations such as the WTO contrast with less successful ones such as the ILO and with a number of remarkably successful international environmental regimes (e.g. Haas et al. 1993; Victor et al. 1998; Young 1999; Miles et al. 2001).

The transformation of an international regime into an organization may largely amount to an act of symbolic policy-making because it does not per se increase governance capacity (see also Charnovitz 2002, 337). The mere fact that the original General Agreement on Tariffs and Trade (GATT), which had evolved over almost 50 years, was converted into the WTO in 1994 did not significantly enhance the governance capacity of the institution. Likewise, environmental governance would not significantly change if the regime for the protection of the global climate or the regime for the protection of the Baltic Sea (Helsinki Commission) were transformed into small international organizations. Hence, *labelling* an existing international regime (agreement) as an organization is largely a symbolic act. We do not want to claim that symbolic politics is irrelevant (Edelman 1967). States may have good reasons to take symbolic action such as the transformation of the Conference for Security and Cooperation in Europe into the Organization for Security and Cooperation in Europe without enhancing the authority of the institution after the end of the cold war to symbolize the new security situation in Europe. From the perspective of cooperation theory, however, symbolic action cannot be expected to systematically enhance the governance capacity of a given institution.

3   The Decision-Making Capacity of International Institutions

An institutional reform will only be relevant for the effectiveness of international environmental governance, if it significantly affects the decision-making capacity of international institutions. In this section, we argue that this factor may be influenced by the design of decision-making processes (subsection one) and by the delimitation of the areas governed (subsection two).

The Design of Decision-Making Processes

International governance institutions constitute decision-making apparatuses that produce collectively binding decisions and supervise their implementation. Active governance intended to change undesired behaviour and improve sub-optimal outcomes requires collective decision-making (see also Keohane 1993; Gehring 2002). As a group, the actors choose how they individually ought to behave in order to bring about a desired cooperative outcome. Hence, collective decision-making constitutes the core of active governance. All international institutions encompass their own decision-making processes (see Levy, Young and Zürn 1995) that enable their members to adapt and develop these institutions dynamically (Gehring 1994).

Simple ad-hoc negotiations are the least demanding mechanism for the making of collective decisions under the 'anarchic' conditions of the international system. This mechanism is frequently resorted to by states in international relations. It allows governments individually to pursue their own distributive interests based on their bargaining power, and collectively to mould norms. It is generally assumed that the distribution of the benefits of an arrangement emerging from a bargaining process will largely reflect the power constellation existing outside the negotiations (Elster 1989, 50-96). Coordination by intergovernmental negotiations is, however, subject to several limitations. First, negotiations are frequently slow and cumbersome. Even if all participants intend to reach agreement, they must distribute cooperation gains and will compete for the biggest possible 'piece of the cake' (Lax and Sebenius 1986). Second, the ability to coordinate by means of simple negotiations decreases with the complexity of the negotiating agenda. If a negotiation addresses many interconnected issues, participants will face increasing difficulties in assessing concessions and proposed deals. These difficulties will be exacerbated, if the issues under negotiation are marked by scientific uncertainty that characterizes many environmental issues. As a consequence, transaction costs of negotiations and the likelihood of failure increase. Third, the dynamic change inherent in many economic, technological and environmental policy areas requires that agreements be flexibly adapted to changing circumstances, which may be difficult to attain in ad-hoc negotiations. Fourth, rational actors can be interested in establishing particularly credible commitments binding their partners and themselves by effective dispute resolution and sanctioning mechanisms, which can hardly be provided for in simple negotiations.

The limitations inherent in coordination by simple negotiations provide incentives for rational actors to devise more ambitious institutional arrangements. They may do so in at least three ways. First, participating actors may postpone some of the issues to be dealt with to later negotiating rounds in order to reduce complexity or to enhance the adaptability of the institution. They may enter into 'incomplete contracts' (Williamson 1985), which regulate only the most pertinent aspects of an issue and postpone everything else to later decisions. For example, parties to the Montreal Protocol of 1987 agreed to periodically review and further develop the phase-out schedule of ozone-depleting substances such as chlorofluorocarbons (CFCs). And the United Nations Framework Convention on Climate Change was intended to provide an institutionalized forum for subsequent negotiations on emission reduction obligations. Actors thus tend to create enduring negotiating processes. Later negotiating rounds inescapably take place within an institutionalized context, which influences the preferences of the participants and makes the development of the institution path-dependent (Young 1994, Pierson 1996). If a considerable number of similar decisions with a limited scope are to be taken over time, general criteria may evolve that guide these decisions, limit the room for manoeuvre in the negotiations and further a transition to an exchange of reasonable arguments (Risse 2000) instead of pure bargaining based on power.

Second, actors may create specialized decision-making processes (e.g. scientific assessments, non-compliance procedures, or simple negotiating sub-groups) that focus on some aspects of the entire decision load and free them from doing everything at the same time within the same process. In this case, the overall package is elaborated within a number of parallel or consecutive negotiation processes with a limited scope that fulfil complementary functions and establish a division of labour. The resulting specialization within country delegations frequently promotes agreement, because it supports recourse to accepted expert knowledge and supports the emergence of 'epistemic communities' (Haas 1992). Even if, as is frequently the case, the sub-agreement forms part of the overall negotiation package that needs to be adopted by consensus ('nothing is agreed until everything is agreed'), the relevance of procedural rules increases and the ability of the participants to resort to their bargaining power outside the negotiations is limited. For example, the scientific, technological and economic assessment panels established within the framework of the Montreal Protocol proved to have a significant influence on the negotiations on the phase out of ozone-depleting substances, although they constituted merely advisory bodies with no formal decision-making powers (Parson 2003). Likewise, international climate negotiations regularly proceed in a number of separate contact groups addressing different items of the negotiating agenda. Specialization of negotiation processes facilitates a transition from interest-based bargaining to an exchange of reasoned arguments (arguing).

Finally, actors may forgive their veto power that would otherwise enable them to block unacceptable decisions by subjecting themselves to majority decision-making or to decisions by committees with limited membership. Delegation of decision-making authority will be possible even under the conditions of the current international system, if the individual decisions are linked so closely that they can only be accepted or rejected in their entirety and a 'selective exit' (Weiler 1991, 2412) is excluded. 'Horizontal' decision-making based upon the consensus of all participating actors can be supplemented by a 'vertical' component, provided that the overall package creates a net benefit for all participants.

The decision-making procedures of several international environmental regimes are surprisingly far-reaching and include delegation of decision-making authority. Member states of the regime for the protection of the ozone layer may adopt binding adjustments of emission reduction commitments by a qualified majority, and an Executive Committee composed of seven developing countries and seven industrialized countries decides on the allocation of financial resources for the phase-out of ozone-depleting substances in the South (DeSombre and Kauffman 1996; Biermann 1997). In the International Whaling Commission and in the international regime on trade in endangered species decisions on catch quota and the listing of endangered species respectively are taken by majority, as are far-reaching decisions in other environmental regimes (Churchill and Ulfstein 2000: 638-41). In an increasing number of international environmental regimes, separate procedures and committees have also been established to identify and address cases of non-compliance. The compliance committee under the Kyoto Protocol even is to decide on significant sanctions (Ehrmann 2000; Churchill and Ulfstein 2000: 643-47; Oberthür and Marr 2002).

While differentiation of negotiation processes increases the autonomy of the institution vis-à-vis its member states and reduces their ability to control decisions, it enhances the decision-making capacity of an international institution. Rational actors will trade off the gains from more flexible decision-making against the (partial) loss of control over the content of institutionally produced decisions (Koremenos et al. 2001). In simple ad-hoc negotiations, the participants constitute the only source of influence on results. If decisions are taken within an established institution, they will be influenced by prior decisions. If they are made in a network of specialized negotiating groups, they may be affected by expert considerations. If collectively binding decisions are taken by a majority or specialized committees with limited membership, they will not require consent by all member states and may not even require consent by any member.

Decision-making arrangements may be improved within existing international environmental agreements without founding a World Environment Organization. Existing arrangements are tailor-made for addressing particular institutions and reflect the willingness of their members to trade off influence against an increased overall decision-making capacity. International environmental regimes develop over time and their arrangements can be adapted to new needs. While there is room for further improvement, progress is frequently precluded by the resistance of some member states based on parochial interests. It is difficult to see how a WEO could systematically contribute to overcoming this resistance.

The Delimitation of Problem Areas

The issue-areas governed by international institutions are not externally given, they are socially constructed by the participants in the process of their interaction (Haas 1975). International negotiations cannot address all cooperation problems pending between states simultaneously, because the multitude of issues would be impossible to handle. Governments can negotiate about almost anything, but not about everything at the same time in the framework of a single negotiating round. Therefore, they identify certain problems to be dealt with in, and exclude others from, a particular negotiating round. The issue-areas governed by international institutions are always 'artificially' delimited. Their boundaries are contingent, i.e. they could have been drawn differently. In bilateral negotiations, the participants can enhance the scope for cooperation by enlarging the issue-area and creating package deals that are composed of partial deals with an asymmetric distribution of benefits (Tollison and Willet 1979). In multilateral negotiations, a deliberate expansion of the agenda is more difficult to achieve, because the number of bilateral relations and of unintended side-effects to be taken into account increase exponentially with the number of participating actors. As a result, the delimitation of multilateral issue-areas usually develops around a substantive core and is frequently subject to protracted pre-negotiations (Gross Stein 1989).

The delimitation of an issue-area has far-reaching consequences for the ensuing negotiation process. Adding or subtracting issues (and parties) will change the constellation of interests within the negotiation process and determine the potential for cooperation (Sebenius 1983). The delimitation of an issue-area establishes effective limits for the attention of the participants (Scharpf 1991) so that actors jointly accept the partition of reality in order to limit the complexity of the respective negotiating agenda. As long as the negotiating partners accept it, an actor does not have an incentive to depart from the definition unilaterally, even if he would have preferred different boundaries. Accordingly, the participating actors will define their preferences in the negotiations with regard to the issues under consideration therein, while all other issues are difficult to introduce and may therefore be ignored for the time being.

Actors participating in negotiations are faced with the challenge to collectively optimize the delimitation of an issue-area so as to enhance the opportunities for successful cooperation and the prospect of achieving a mutually accepted outcome. An optimal delimitation of an issue-area will avoid two pathologies. A very narrow definition promises a manageable scope of negotiations, but involves the risk of providing too little potential for cooperation and trade-offs (Whalley and Zissimos 2001; 2002). In contrast, a very broad definition will provide ample room for linkages and trade-offs between actors, but creates the risk that the negotiating partners are overwhelmed by the complexity of too many problems. The suitable scope of an issue-area depends upon the particularities of the issues at stake. Some issue-areas are comparatively broadly defined in order to facilitate the emergence of numerous small-scale cooperation projects which might not have been born, or which would not survive, separately. The International Labour Organization (ILO), for example, has helped adopt several hundred conventions defining minimum social and labour standards (Cruz et al. 1996) for which states might not have gathered and attended separate conferences and established independent supervisory mechanisms. The same is true for the World Health Organization which addresses a broad range of issues. In these cases, transaction costs may be assumed to be high as compared to cooperation gains.

In other cases, broad-range international institutions have increased the prospects of beneficial trade-offs. International trade, for example, is based on the systematic exploitation of comparative advantages. A country will specialize in the production and export of those products in which it is competitive, whereas it will import those products which it cannot produce competitively. Despite some intra-industry trade, liberalizing just trade in cars would not lend itself to international cooperation because producers and exporters of cars would benefit from this step, but not the importing countries. The integration of different product markets creates significantly more room for mutually beneficial cooperation. The same is true for the linkage of trade in goods (GATT) with trade in services (GATS) and the protection of intellectual property (TRIPs). Hence, international trade is regulated by a single international institution, namely WTO (Hoekman and Kostecki 1995), and not by a set of separate product regimes. The striking institutional fragmentation of international environmental politics reflects the high political salience of environmental issues and their particular problem structure. Unlike trade policy, environmental cooperation projects do not require integration because most international and global environmental issues address the protection of common goods and encompass a prisoners' dilemma constellation of interests. They can be regulated in separate cooperative arrangements that ensure mutual benefits for the members. For example, all members of the international ozone regime or the regime on biological diversity benefit from an enhanced protection of the ozone layer and an improved conservation of nature (in the case of developing countries partially ensured by side-payments).

Unlike international health and labour protection policies, environmental cooperation projects are usually sufficiently important to be institutionalized separately from existing institutions. Transaction costs involved in launching new cooperative projects are reduced by a number of existing international organizations or semi-independent organizational structures. Existing organizations such as UNEP (protection of the ozone layer, transboundary movements of hazardous wastes), the United Nations Economic Commission for Europe (UNECE; long-range transboundary air pollution in Europe), the UN General Assembly (climate change), and IMO (pollution from ships) have supported the emergence of most international environmental regimes. Compared to the huge investments and far-reaching changes of administrative procedures required to implement international environmental commitments, transaction costs related to the maintenance of separate institutions are small. Hence, even international environmental regimes created within the framework of international organizations such as those mentioned above regularly gain institutional autonomy from their parent institutions (Churchill and Ulfstein 2000).

The institutional fragmentation of international environmental governance indicates the strength rather than the weakness of environmental cooperation. While the delimitation of issue-areas governed by international institutions is subject to design efforts by the actors involved, it is not entirely incidental. In general, issue-areas are to ensure mutual benefits for the participants in order to enable effective governance. A wrong delimitation of an issue-area can render cooperation impossible. The multitude of well-functioning environmental institutions indicates that actors have, for the most part, succeeded in defining viable issue-areas in international environmental governance and that an integration of issue-areas is not required in order to ensure mutual benefits of the parties involved. In subsection two of Section 4, we will turn in more detail to the question of whether an integration of issue-areas may nevertheless have the potential of enhancing international environmental governance.

4   Options for a World Environment Organization: Three Models

The idea of a more encompassing WEO is closely related to the integration of environmental issue-areas. Neither the mere transformation of individual regimes into single-issue organizations nor the re-design of their decision-making procedures correspond to the idea of a WEO and are thus considered here. The organization is generally expected to form the institutional core of international environmental governance. Consequently, it would have to provide the home for a number of international environmental regimes such as the regimes on global climate change, for the protection of the ozone layer and for the preservation of biodiversity.

In this section, we develop three basic models for a future WEO. While an almost unlimited number of options exist, the specific design of a WEO will follow one of these models (for similar distinctions see Biermann 2000 and 2001; German Advisory Council on Global Change 2001). The models differ with respect to the two parameters discussed in the preceding section, namely the decision-making processes and the delimitation of the issue-area(s) governed. Actors could create a formal umbrella organization without changing issue-areas and decision-making procedures of existing regimes substantively (see *UN Model*). They could also integrate existing issue-areas more substantively without introducing significantly different decision-making procedures (see *WTO Model*). Finally, they could both integrate issue-areas and fundamentally re-organize decision-making therein (see *EU Model*).

UN Model: An Umbrella Organization

States could establish a World Environment Organization limited to providing a formal umbrella for existing sector-specific environmental regimes. It would leave substantively untouched the current institutional structure of international environmental governance. The established boundaries of the issue-areas governed by international regimes and their existing decision-making procedures would remain unchanged. The organization could stimulate international environmental cooperation by lowering the threshold of regime building and reduce transaction costs, e.g. by offering constant secretariat services, or fulfilling certain auxiliary functions. Many proposals for establishing a WEO (Esty and Ivanova 2002; Biermann 2000 and this volume) emphasize these aspects, that are also the major focus of proposals for 'clustering' multilateral environmental agreements (Oberthür 2002; Moltke in this volume). A WEO designed in this way would follow the model of the United Nations which also provides a comparatively loose umbrella for a number of rather independent regulatory activities in separate issue-areas such as human rights or the law of the sea (Mingst and Karns 2000; White 2002).

A WEO following the UN model would not significantly affect the governance capacity of institutions making international environmental policy. The currently separate environmental issue-areas would not be integrated, because the sector-specific decision-making processes would remain in place. The participating actors would continue to determine their preferences in relation to those issues falling inside the respective issue-areas, while ignoring other issues. Opportunities for cooperation would continue to arise exclusively as a result of these sector-specific preferences. If decision-making proceeded separately for each issue-area—although within the framework of an umbrella organization—negotiators would not receive additional incentives to coordinate their sector-specific activities and to look for possible issue-linkages or for package deals cutting across the boundaries of established issue-areas. Those negotiating climate change would continue to focus on measures to stabilize the global climate, while members of the regime on biological diversity would continue to concentrate on preserving biodiversity. Whereas an exchange of information may be facilitated, resulting tensions between both regimes regarding forestry activities (maximization of carbon sequestration versus conservation of biological diversity) would persist. Likewise, the mechanisms for supervising and facilitating implementation such as non-compliance procedures and other functional bodies would not significantly change, because they would remain sectorally organized.

A WEO constructed after the UN model could be expected to realize limited efficiency gains at best, but it would not make a significant contribution to the solution of problems of international environmental governance related to decision-making, implementation and coordination.

A limited potential for combining certain auxiliary functions of environmental regimes (e.g. reporting, review of implementation) might exist, but gains would be moderate (Oberthür 2002). The bigger problems of international environmental governance could not be solved because this WEO would not significantly change the delimitation of existing issue-areas or the design of the related decision-making processes. The creation of an umbrella organization would thus largely be a matter of symbolic politics.

WTO Model: Creating Comprehensive World Environment Rounds

Alternatively, a WEO could systematically integrate issue-areas so far governed by separate international environmental regimes without abandoning the familiar intergovernmental structure. Issue-areas will be integrated, if the specific decision-making processes of existing and future international environmental regimes, in particular their annual or biennial conferences of the parties, are merged into more encompassing negotiating rounds. International environmental policy would then be developed in recurring 'global environmental rounds'. This design follows the approach of the WTO, because the regulation of world trade is developed in comprehensive world trade rounds (Hoekman and Kostecki 1995: 12-20).

In contrast to the UN model, such a WEO would have a substantial impact on international environmental governance. Members would determine their preferences related to a broader negotiating agenda taking into account additional options for package deals and trade-offs. They might also more easily detect, and might attempt to avoid, negative side-effects of a given regulation on other policies pursued within the same institution. However, these potentially positive effects on international environmental governance contrast with important drawbacks. The increased potential for issue-linkage resulting from the integration of issue-areas is unlikely to help advance international environmental governance. Whalley and Zissimos (2001 and 2002) rightly insist that increased opportunities for issue-linkages and side-payments facilitate agreement in international negotiations. However, issue-linkage will only be helpful, if asymmetries in the distribution of costs and benefits are complementary across issues so that different actors benefit (most) from cooperation on different subjects and all actors equally benefit from the overall package. Unfortunately, complementary interests, which characterize trade negotiations, do not systematically exist across different issue-areas in international environmental policy. For example, it is unlikely that difficulties in the international cooperation to combat climate change would be more easily overcome, if negotiations were combined with those on ozone depletion, biodiversity or other regimes. Complementarity of interests does not exist because the United States is currently the laggard in many, if not most, global environmental issues (Paarlberg 1999). The United States can hardly be expected to accept stringent controls on greenhouse gas emissions in order to ensure an effective protection of, for example, the ozone layer and cooperation on persistent organic pollutants. If this situation gave way to a complementarity of interests, it would be incidental and temporary and could not be expected to provide a firm foundation for long-term cooperation.

Benefits from an integration of issue-areas are limited because international environmental governance is predominantly about the preservation of collective goods rather than club goods. Free international trade has the properties of a club good that is accessible only to the members of the club (Cornes and Sandler 1999). States are effectively excluded from reaping the benefits of a liberalized world trade unless they open their own markets (Hoekman and Kostecki 1995, 27-30). In contrast, environmental protection is frequently a collective good. It will be difficult to prevent a state from taking a free ride if it cannot be excluded from enjoying the collective good of environmental protection. Countries refusing to cooperate to protect the ozone layer cannot be excluded from the benefits of a stabilized ozone layer. Accordingly, states have an incentive to stay out of costly cooperation (Olson 1965) that will increase with every issue that a country opposes. Thus, a WEO following the WTO model threatens to undermine its own basis and endangers gains so far realized through sector-specific cooperation in international regimes.

Likewise, issue-linkage through integration of issue-areas does not help pressure non-cooperating states and enforce implementation of international environmental commitments. Proponents of environmental protection cannot credibly threaten to make protection of the ozone layer conditional on United States acceptance of controls on greenhouse gases, because realizing this threat would harm themselves at least as much as the opponent. The same logic applies to the enforcement of obligations. While disregard of obligations within WTO may be effectively prosecuted by excluding non-complying countries from benefits in any suitable area of international trade, this threat is usually not available in environmental institutions: A country's non-compliance with obligations to conserve biological diversity cannot usefully be responded to by not complying with commitments to protect the ozone layer. Modest additional opportunities for issue-linkages and side-payments contrast with a significantly enhanced complexity of negotiations, which makes it more difficult to reach agreement. If states had unlimited information processing capacity, they would best deal with all problems pending among them at the same time. In reality, complexity creates significant impediments for decision-making and reaching agreement, because more issues are to be dealt with in a single negotiation process. Experience from available precedents suggests that complexity is a factor that seriously limits effective decision-making and significantly slows down the process. The Uruguay Round of trade negotiations took some eight years (1986-1994: Hoekman and Kostecki 1995: 19-20), while the negotiations at the Third UN Conference on the Law of the Sea lasted even for nine years (1973-1982: Sebenius 1984).

In several respects, a WTO-like WEO does not change the status quo at all. It is unlikely that it is apt to mobilize the additional financial resources needed to reinforce the capacity of developing countries to implement international obligations and develop effective environmental policies. There is no indication that industrialized countries might be more willing to provide additional financial resources to assist implementation of international environmental commitments in developing countries if issue-areas were integrated. Why should they be prepared to do so only because the ozone regime has, for example, been merged with the climate change regime and become part of a larger institutional complex? It should also be noted that the Global Environment Facility (GEF) has already been established as an overarching institution providing financial assistance supporting the implementation of several international environmental agreements (Oberthür 2002, 324-25).

A WTO-like WEO would also be unlikely to help resolve the coordination problems existing between environmental and economic institutions such as the WTO (Brack 2002). Several proposals for a WEO are focusing on the interface between economic and environmental governance (e.g. Runge 2001; Charnovitz in this volume). However, it is difficult to see how a larger environmental organization could ensure that agreements between WTO members do not undercut environmental regulation. The difference in power between the WTO and international environmental governance will continue irrespective of the existence of a WEO because the WTO can grant *and withdraw* trade advantages, while any environmental institution supplying a collective good cannot exclude individual actors from benefiting (see above). Where particular opportunities for enforcement exist, they can be used without establishing a WEO. Side-payments and technical assistance can be withdrawn, with the significant drawback that such action only affects the poorer members and diminishes their willingness to cooperate. More importantly, several multilateral environmental agreements, including the Montreal Protocol for the protection of the ozone layer and the 1973 Convention on International Trade in Endangered Species of Wild Fauna and Flora (CITES), allow for authorizing or imposing specific trade sanctions in response to non-compliance (Charnovitz 1996). Pressure on the WTO (and other economic institutions) will largely depend on the determination of the members of these regimes to effectively implement such sanctions. It seems that WTO, through the jurisdiction of its Dispute Settlement Body, increasingly accepts appropriately designed environmental trade sanctions (Charnovitz 1998) so that, in practice, trade sanctions (and other enforcement options) can already be employed by the parties to environmental regimes. It is difficult to see how the establishment of a WEO could add significant new opportunities in this respect.

Finally, a WTO-like WEO could make a difference for coordination problems between the environmental institutions integrated into the WEO, but the benefits to be reaped would remain very limited. A WEO would facilitate exchange of information and coordination across environmental issue-areas contributing to a better integration of approaches. However, recent research suggests that tensions between environmental institutions are relatively rare and have been handled relatively successfully within the current fragmented system of environmental institutions (Oberthür and Gehring 2003). The benefits to be realized from a better integration of environmental issue-areas are therefore moderate at best. Altogether, it is highly questionable whether the benefits that may be reaped in this area justify the costs and dangers of increased complexity and unproductive issue linkage. While a WEO following the WTO model would significantly change environmental policy-making, it does not promise significant progress towards the resolution of the major problems of international environmental governance. To the contrary, it is likely to be dysfunctional, because it creates disincentives rather than incentives for accepting additional environmental commitments and threatens to overwhelm negotiators with an undesirable complexity of issues that would retard action.

EU Model: Delegating Extensive Competencies to a WEO

To avoid the negative consequences of the WTO model, a WEO could both integrate the issue-areas of existing environmental regimes and systematically reorganize the related decision-making processes. It would then be shaped after the European Union (EU). The EU constitutes a single encompassing institution governing an extensive area comprising numerous policies. In principle, it thus provides extensive opportunities for linking issues and facilitating cooperation by side-payments. Nevertheless, decision-making is not overwhelmed by an overly complex agenda of issues because opportunities for linkage are severely limited in practice. Despite grand bargains on the European Treaties (Moravcsik 1998), there are no comprehensive negotiation rounds. Instead, decisions are largely delegated to subsidiary decision-making processes, supranational organs (European Commission, European Court of Justice) and independent bodies or agencies (Majone 1997). Each of these processes is specialized on some issues, so that complexity is reduced. Their decision-making capacity is enhanced because they regularly employ majority decision-making, and in many cases member states do not play the dominant role any more. Moreover, decisions are not part of large packages to be agreed upon at the end, but enter into force separately. In order to avoid that states only adhere to agreements that are to their liking, the option of 'selective exit' (Weiler 1991) is firmly closed, and member states must either accept all commitments of the institution or sacrifice their membership.

A WEO following the European Union model would promise to make a substantial contribution to resolving the main problems of international environmental governance (Pollack 2003). First, majority voting and delegation of decision-making authority would reduce the decision-making problems notorious in international institutions. It would not only prevent individual actors from blocking decisions, but produce decisions that are less influenced by the parochial interests of individual states. Second, the institutional design would require, and thus allow, that member states establish powerful supervisory and enforcement mechanisms, because the institution would comprise numerous decisions with asymmetrical distributional effects that would not have gained the support by all members concerned. Otherwise, effective implementation of commitments could not be ensured. Third, a 'supranational' WEO would allow for the establishment of overarching criteria and collision rules to be followed in the subordinated specialized decision-making processes in order to resolve coordination problems and potential conflicts such as that between the Kyoto Protocol and the Convention on Biological Diversity.

Even the governance capacity of a WEO designed according to the European Union model would have notable limits. The attractiveness and power of the EU stem primarily from the advantages offered by the single market as opposed to the prospect of participating in European environmental governance. The same is true for the North American Free Trade Agreement (NAFTA) and the accompanying North American Agreement on Environmental Cooperation (NAAEC) and its Commission for Environmental Cooperation (Runge 2001). A similar link to other policy fields would presumably not be part of a newly established WEO because it would transform the concept of a World *Environment* Organization into something much broader. A WEO integrating only environmental issue-areas could not be expected to resolve the enforcement problems encountered with respect to collective environmental goods referred to above. Even the most stringent supervisory and enforcement mechanism could not prevent non-compliant countries from taking a free ride on members by leaving the organization or ignoring its rules. The notable strength of EU enforcement of environmental regulation is in particular due to the linkage with other policies such as single market policy and agricultural and regional policies, rather than the particular design of environmental policy making. For the same reason, such a WEO could not be expected to mitigate tensions with economic institutions such as the WTO.

Finally, even the most powerful 'supranational' WEO is unlikely to raise substantial additional financial resources. There is no reason to assume that the integration of issue-areas and the reorganization of decision-making processes would increase the preparedness of member states to provide additional contributions. Paying EU member states have not been willing to agree to and maintain a massive transfer of financial resources in the framework of the EU Structural Funds and the EU Common Agricultural Policy because of the desirability of the goals pursued thereby, but because of the substantial benefits they reap from European integration in other areas (e.g. single market, monetary union). Similar trade-offs beyond the field of the environment are not in sight for a WEO. Most important, a 'supranational' WEO relying on the European Union model cannot be expected to be realized in the foreseeable future, because this would require the transfer of far-reaching competencies to an international organization. Supranationalism has been confined to the European Union so far where it depends on a comparatively high degree of political, economic and social coherence in a single region of the world. We cannot see any indication that a significant number of states from different regions would be prepared to consider ceding sovereignty to an international institution to the extent required.

5   Conclusion

The establishment of a world environment organization does not promise to enhance international environmental governance. Currently, this governance occurs through hundreds of separate international regimes and is characterized by a high degree of institutional fragmentation. Proponents of a WEO aim at integrating existing regimes into a more centralized institution in order to improve decision-making, implementation and coordination in international environmental governance. However, they have largely failed to answer the fundamental question of how and why governance within the framework of an organization would be superior to the status quo. From an institutionalist perspective, proposals for establishing a WEO lack promise because such a WEO cannot at the same time be realistic, significant and beneficial for international environmental governance.

Any *institutional* reform can significantly affect international environmental governance only if it succeeds in modifying the design of the decision-making processes applied in international environmental institutions and the scope of the issue-areas covered by these institutions.

International institutions consist of systems of rules and norms that are established and developed by their members in order to govern distinct issue-areas. The design of their decision-making procedures and the scope of their issue-areas are important determinants of their governance capacity. Replacing the consensus principle currently prevailing in international relations at least partially with majority voting and delegation of decision-making authority to limited-membership bodies and independent agencies could substantially enhance the ability to arrive at decisions in international institutions. An issue-area's ideal scope will be reached if it allows for sufficient trade-offs between issues so as to ensure net benefits for all major actors, while preventing negotiations from becoming over-complex and unmanageable. From an institutionalist perspective, any significant effect of an institutional reform of international environmental governance will be based on modifications either of decision-making processes or of boundaries of issue-areas or of both. Such institutional reform can help strengthen environmental interests but it cannot substitute for a lack of political interest in and support for environmental protection.

The establishment of an organization as such does not promise to significantly improve international environmental governance. International organizations possessing legal personality and bureaucracies do not a priori possess a higher governance capacity than international regimes, which are based upon one or several international treaties short of possessing legal personality. Neither are international organizations throughout characterized by particularly far-reaching decision-making procedures, nor is the construction of the issues-areas governed necessarily superior to international regimes. Both organizations and regimes delegate decision-making authority and employ majority decision-making to varying extents, with sophisticated decision making within international environmental regimes reaching far beyond arrangements found in many international organizations. Both of them can govern smaller or larger problem areas depending on whether the respective area allows for the emergence of international cooperation. Finally, there are both examples of more and less effective international organizations and international regimes with many environmental regimes having found to be particularly successful.

Depending on its design, a newly established WEO could either constitute symbolic action, or create a host of negative side-effects, or be unrealistic. A WEO designed according to the UN model would assemble the existing environmental regimes under a common roof without affecting the scope of their issue-areas or their decision-making procedures. Creating this umbrella organization would be a symbolic act that would have little or no effect on governance. In contrast, a WEO following the WTO model would profoundly change decision-making by systematically integrating existing environmental issue-areas and leading to comprehensive 'world environment rounds'. As a consequence, however, the complexity of negotiation rounds would increase, while opponents of parts of the comprehensive package would face incentives to stay outside the institution or ignore part of its regulations. A WEO relying upon the European Union model could be expected to successfully avoid the complexity trap. Members would be requested to accept majority decision-making and delegation of decision-making authority and would be faced with the choice of either accepting all or none of the decisions taken within the WEO. While this WEO would have some prospect of contributing to solving the problems of international environmental governance, it would require a far-reaching transfer of competencies from nation states to the organization. Since such a transfer is currently widely unacceptable for states in the international system, an EU-like WEO can be considered grossly utopian.

The effectiveness of any WEO, however designed, would remain limited by the particular problem structure prevailing in international environmental governance. A WEO would hardly be able to prevent countries from taking a free ride on faithful members by staying outside the organization or ignoring their obligations. It also cannot be expected per se to enhance the preparedness of member states to provide increased financial funds for environmental protection. There are ample opportunities for improving the current institutional arrangements of international environmental governance. Efforts to enhance the effectiveness of new and emerging sectoral environmental regimes may be complemented by endeavours aiming at improved efficiency and coherence by integrating secretariat services and meetings of parties and strengthening the catalytic and facilitative role of UNEP short of establishing a WEO. From an institutionalist perspective, environmental protection will be better served if the political resources available are invested in achieving progress in the development of the existing institutional arrangements.

References

Abbott, K.W. and D. Snidal (1998), 'Why States Act through Formal International Organizations', *Journal of Conflict Resolution*, vol. 42 (1), 3–32.

Barnett, M.N. and M. Finnemore (1999), 'The Politics, Power and Pathologies of International Organizations', *International Organization*, vol. 53 (4), 699–732.

Bauer, S. and F. Biermann (2005), 'The Debate on a World Environment Organization: An Introduction', in F. Biermann and S. Bauer (eds), *A World Environment Organization. Solution or Threat for Effective International Environmental Governance?*, Ashgate, Aldershot, 1–23.

Beisheim, M., S. Dreher, G. Walter, B. Zangl and M. Zürn. (1999), *Im Zeitalter der Globalisierung? Thesen und Daten zur gesellschaftlichen und politischen Denationalisierung*, Nomos, Baden-Baden.

Biermann, F. (1997), 'Financing Environmental Policies in the South. Experiences from the Multilateral Ozone Fund', *International Environmental Affairs*, vol. 9 (3), 179–218.

Biermann, F. (2000), 'The Case for a World Environment Organization', *Environment*, vol. 42 (9), 22–31.

Biermann, F. (2001), 'The Emerging Debate on the Need for a World Environment Organization: A Commentary', *Global Environmental Politics*, vol. 1 (1), 45–55.

Biermann, F. (2005), 'The Rationale for a World Environment Organization', in F. Biermann and S. Bauer (eds), *A World Environment Organization. Solution or Threat for Effective International Environmental Governance?*, Ashgate, Aldershot, 117–144.

Brack, D. (2002), 'Environmental Treaties and Trade: Multilateral Environmental Agreements and the Multilateral Trading System', in G.P. Sampson and W.B. Chambers (eds), *Trade, Environment and the Millennium*, Second Edition, UN University Press, Tokyo, 321–52.

Charnovitz, S. (1996), 'Trade Measures and the Design of International Regimes', *Journal of Environment and Development*, vol. 5 (2), 168–196.

Charnovitz, S. (1998), 'The World Trade Organization and the Environment', *Yearbook of International Environmental Law*, vol. 8, 98–116.

Charnovitz, S. (2002), 'A World Environment Organization', *Columbia Journal of Environmental Law*, vol. 27 (2), 321–57.

Charnovitz, S. (2005), 'Toward a World Environment Organization: Reflections upon a Vital Debate', in F. Biermann and S. Bauer (eds), *A World Environment Organization. Solution or Threat for Effective International Environmental Governance?*, Ashgate, Aldershot, 87–115.

Churchill, R. and G. Ulfstein (2000), 'Autonomous Institutional Arrangements in Multilateral Environmental Agreements: A Little-Noticed Phenomenon in International Law', *The American Journal of International Law*, vol. 94, 623–59.

Cornes, R. and T. Sandler (1999), *The Theory of Externalities, Public Goods and Club Goods*, 2nd ed., Cambridge University Press, Cambridge (UK).

Cruz, H.B. de la, G. Potobsky and L. Swepston (1996), *The International Labor Organization. The International Standards System and Basic Human Rights*, Boulder (Col.).

DeSombre, E. R. and J. Kauffman (1996), 'The Montreal Protocol Multilateral Fund: Partial Success Story', in R.O. Keohane and M.A. Levy (eds), *Institutions for Environmental Aid: Pitfalls and Promise*, MIT Press, Cambridge (Mass.), 89–126.

Edelman, M. (1967), *The Symbolic Uses of Politics*, University of Illinois Press, Urbana/ Chicago/London.

Ehrmann, M. (2000), *Erfüllungskontrolle im Umweltvölkerrecht—Verfahren der Erfüllungskontrolle in der umweltvölkerrechtlichen Vertragspraxis*. Nomos, Baden-Baden.

Elster, J. (1989), *The Cement of Society. A Study of Social Order*, Cambridge.

Esty, D.C. (1994), *Greening the GATT. Trade, Environment and the Future*, Harlow.

Esty, D.C. and M. Ivanova (2001), *Making International Environmental Agreements Work: The Case for a Global Environmental Organization*, Yale Center for Environmental Law and Policy Working Paper Series, Working Paper 2/1, New Haven (Conn.), May 2001.

Esty, D.C. and M. Ivanova (2002), 'Revitalizing Global Environmental Governance: A Function-Driven Approach', in D. C. Esty and M. Ivanova (eds), *Global Environmental Governance: Options and Opportunities*. New Haven, CT: Yale School of Forestry and Environmental Studies, 181–204.

Gehring, T. (1994), *Dynamic International Regimes: Institutions for International Environmental Governance*, Peter Lang, Frankfurt/Main.

Gehring, T. (2002), *Die Europäische Union als komplexe internationale Organisation. Wie durch Kommunikation und Entscheidung soziale Ordnung entsteht*, Nomos, Baden-Baden.

German Advisory Council on Global Change (2001), *New Structures for Global Environmental Policy*, Earthscan, London.

Gross Stein, J. (1989), 'Getting to the Table: The Triggers, Stages, Functions and Consequences of Pre-negotiation', in J. Gross Stein (ed.), *Getting to the Table: The Processes of International Pre-negotiation*, Baltimore, 239–68.

Haas, E.B. (1975), 'Is There a Hole in the Whole? Knowledge, Technology, Interdependence and the Construction of International Regimes', *International Organization*, vol. 29 (3), 827–876.

Haas, P.M. (1992), 'Banning Chlorofluorocarbons: Epistemic Community Efforts to Protect Stratospheric Ozone', *International Organization*, vol. 46 (1), 187–224.

Haas, P.M., R.O. Keohane and M. Levy (1993) (eds), *Institutions for the Earth: Sources of Effective International Environmental Protection*, MIT Press, London.

Hoekman, B.M. and M.M. Kostecki (1995), *The Political Economy of the World Trading System: From GATT to WTO*, Oxford.

Jackson, J.H. (1999), *The World Trading System. Law and Policy of International Economic Relations*, 2nd ed., Cambridge (Mass.).

Jacquemont, F. and A. Caparrós (2002), 'The Convention on Biological Diversity and the Climate Change Convention 10 Years After Rio: Towards a Synergy of the Two Regimes?', *Review of European Community and International Environmental Law (RECIEL)*, vol. 11 (2), 139–80.

Keohane, R.O. (1989), *International Institutions and State Power. Essays in International Relations Theory*. Westview Press, Boulder (Col.).

Keohane, R.O. (1993), 'The Analysis of International Regimes: Towards a European-American Research Programme', in V. Rittberger (ed.), *Regime Theory and International Relations*, Oxford.

Kirton, J. (2005), 'Generating Effective Global Environmental Governance: The North's Need for a WEO', in F. Biermann and S. Bauer (eds), *A World Environment Organization. Solution or Threat for Effective International Environmental Governance?*, Ashgate, Aldershot, 145–172.

Koremenos, B., C. Lipson and D. Snidal (2001), 'The Rational Design of International Institutions', *International Organization*, vol. 55 (4), 761–799.

Lax, D.A. and J.K. Sebenius (1986), *The Manager as Negotiator. Bargaining for Cooperation and Competitive Gain*, New York.

Levy, M.A., O.R. Young and M. Zürn (1995), 'The Study of International Regimes'. *European Journal of International Relations*, vol. 1, 267–330.

Majone, G. (1997), 'The New European Agencies. Regulation by Information', *Journal of European Public Policy*, vol. 4 (2), 262–75.

Miles, E.L., A. Underdal, S. Andresen, K. Lee, J.B. Skjærseth and J. Wettestad (2001), *Explaining Regime Effectiveness: Confronting Theory with Evidence*, MIT Press, Cambridge (Mass.).

Mingst, K.A. and M.P. Karns (2000), *United Nations in the Post-Cold War Era*, Westview Press, Boulder (Col.).

Moravcsik, A. (1998), *The Choice for Europe. Social Purpose and State Power from Messina to Maastricht,* Cornell University Press, Ithaca NY.

Najam, A. (2003), 'The Case against a New International Environmental Organization', *Global Governance,* vol. 9(3), 367–384.

Najam, A. (2005), 'Neither Necessary, Nor Sufficient: Why Organizational Tinkering Won't Improve Environmental Governance', in F. Biermann and S. Bauer (eds), *A World Environment Organization. Solution or Threat for Effective International Environmental Governance?,* Ashgate, Aldershot, 235–256.

Oberthür, S. (2002), 'Clustering of Multilateral Environmental Agreements: Potentials and Limitations', *International Environmental Agreements: Politics, Law and Economics,* vol. 2, 317–40.

Oberthür, S. and S. Marr (2002), ,Das System der Erfüllungskontrolle des Kyoto-Protokolls: Ein Schritt zur wirksamen Durchsetzung im Umweltvölkerrecht', *Zeitschrift für Umweltrecht,* vol. 13 (2), 81–89.

Oberthür, S. and T. Gehring (2003), *Institutional Interaction: Toward a Systematic Analysis,* paper presented at the 2003 International Studies Association Annual Convention, Portland, 26 February–1 March 2003.

Olson, M. (1965), *The Logic of Collective Action. Public Goods and the Theory of Groups,* Cambridge (Mass.).

Paarlberg, R.L. (1999), 'Lapsed Leadership: U.S. International Environmental Policy Since Rio', in N.J. Vig and R.S. Axelrod (eds), *The Global Environment: Institutions, Law and Policy,* Congressional Quarterly Press, Washington DC, 236–55.

Parson, E.A. (2003), *Protecting the Ozone Layer: Science, Strategy and Negotiation in the Shaping of a Global Environmental Regime,* Oxford University Press, Oxford.

Pierson, P. (1996), 'The Path to European Integration. A Historical Institutionalist Analysis', *Comparative Political Studies,* vol. 29 (2), 123–63.

Pollack, M. (2003), *The Engines of European Integration. Delegation, Agency and Agenda Setting in the EU,* Oxford University Press, Oxford.

Pontecorvo, C.M. (1999), 'Interdependence Between Global Environmental Regimes: The Kyoto Protocol on Climate Change and Forest Protection', *Zeitschrift für ausländisches öffentliches Recht und Völkerrecht,* vol. 59 (3), 709–749.

Risse, T. (2000), '"Let's Argue!" Communicative Action in World Politics', *International Organization,* vol. 54 (1), 1–39.

Runge, C.F. (2001), 'A Global Environment Organization (GEO) and the World Trading System', *Journal of World Trade,* vol. 35(4), 399–426.

Sand, P.H. (ed.) (1992), *The Effectiveness of International Environmental Agreements. A Survey of Existing Legal Instruments,* Cambridge.

Scharpf, F.W. (1991), 'Games Real Actors Could Play. The Challenge of Complexity', *Journal of Theoretical Politics*, vol. 3 (3), 277–304.

Sebenius, J.K. (1983), 'Negotiation Arithmetics: Adding and Subtracting Issues and Parties', *International Organization*, vol. 37 (2), 281–316.

Sebenius, J.K. (1984), *Negotiating the Law of the Sea*, Cambridge (Mass.).

Simmons, B.A. and L.L. Martin (2002), 'International Organizations and Institutions', in W. Carlsnaes, T. Risse and B.A. Simmons (eds), *Handbook of International Relations*, London, Sage, 192–211.

Tollison, R.D. and T.D. Willett (1979), 'An Economic Theory of Mutually Advantageous Issue Linkages in International Negotiations', *International Organization*, vol. 33 (4), 425–49.

UNEP [United Nations Environment Programme] (2002), *Global Environment Outlook GEO-3*, Earthscan, London.

Victor, D.G., K. Raustiala and E.B. Skolnikoff (1998) (eds), *The Implementation and Effectiveness of International Environmental Commitments: Theory and Practice*, MIT Press, Cambridge (Mass.).

von Moltke, K. (2005), 'Clustering International Environmental Agreements as an Alternative to a World Environment Organization', in F. Biermann and S. Bauer (eds), *A World Environment Organization. Solution or Threat for Effective International Environmental Governance?*, Ashgate, Aldershot, 175–204.

Weiler, J.H.H. (1991), 'The Transformation of Europe', *Yale Law Journal*, vol. 100 (8), 2403–83.

Whalley, J. and B. Zissimos (2001), 'What Could a World Environmental Organization Do?', *Global Environmental Politics*, vol. 1 (1), 29–44.

Whalley, J. and B. Zissimos (2002), 'Making Environmental Deals: The Economic Case for a World Environment Organization', in D.C. Esty and M. Ivanova (eds), *Global Environmental Governance: Options and Opportunities*, Yale School of Forestry and Environmental Studies, New Haven, CT, 163–180.

White, N.D. (2002), *The United Nations System. Toward International Justice*, Lynne Rienner, Boulder/London.

Williamson, O.E. (1985), *The Economic Institutions of Capitalism. Firms, Markets, Relational Contracting*, Free Press, New York.

Young, O.R. (1994), *International Governance. Protecting the Environment in a Stateless Society*, Ithaca.

Young, O.R., (1999) (ed.), *The Effectiveness of International Environmental Regimes: Causal Connections and Behavioural Mechanisms*, MIT Press, Cambridge (Mass.).

Chapter 9

Neither Necessary, Nor Sufficient: Why Organizational Tinkering Will Not Improve Environmental Governance

Adil Najam

1   Introduction

There is much in the earlier chapters of this volume that one instinctively agrees with. If one lived in a perfect world one might well have thought about investing resources and time into constructing the perfect environmental organization—but only after the multitude of the far more pressing problems of global environmental governance had been taken care of. This book is itself a testimony to the fact that we do not live in a perfect world. Moreover, there still exist too many problems that demand much more immediate attention. Until that perfect world arrives, talk of organizational tinkering is likely to be a distraction from, and possibly an impediment to, more effective international environmental governance. The concern is not merely that organizational tinkering is not the most pressing priority now. It is also that in the absence of other structural reform, no amount of organizational tinkering is likely to succeed. Such tinkering is neither necessary nor sufficient for better environmental governance. It could easily make things worse by further burdening an already over-burdened system.

The premise of this chapter[1] is that the current debate about global environmental governance with its dominant focus on establishing a super-organization for the environment, represents a serious misdiagnosis of the issues, is unfair to the United Nations Environment Programme (UNEP), and is likely to distract attention from other more important challenges of global environmental governance.

This is not to suggest that there is no 'crisis' of global environmental governance. Earlier chapters in this book have made a convincing case that there is. The crisis, however, is one of 'governance'; of which, organizational structure is but one element and, in this case, a relatively small element (Young 1997). By co-opting the larger discussions on global environmental governance, the discourse on organizational restructuring—under whatever grandiose name such proposals are advertised—are distracting from the more important and immediate challenges of global environmental governance that we face as the Rio compact on environment and development crumbles around us. The thought that any of the competing plans for a WEO or a GEO (World, or Global, Environmental Organization respectively) that are being offered (Runge et al. 1994, Esty 1994, Charnovitz 1995, Biermann 2000, Downie and Levy 2000, Whalley and Zissimos 2001) might actually be taken seriously by the world's governments—as it sometimes seems possible—is even more disturbing. Not only do they show very little promise of actually doing much good to the cause of improved global environmental governance, but also some could actually do harm by distracting international attention from more pressing issues.

It is not the purpose of this chapter to re-examine, or criticize, the details of different schemes for organizational restructuring. Critiques are available elsewhere in the larger literature (Agarwal et al. 1999, Juma 2000, von Moltke 2001a, Newell 2001). Moreover, to do so would be to cede to the premise on which such proposals are based and it is that very premise

---

[1]   This chapter is based on Najam (2003a).

that I wish to question. It should be noted, however, that there is a certain variety in the proposals—ranging from Esty's (1996) GEO that would focus only on 'global' issues, to Biermann's (2000) WEO which would also incorporate more local concerns, to Whalley and Zissimos' (2001) desire to create a 'global bargaining-based entity', to Downie and Levy's (2000) notion of a 'super-UNEP'. However, all such schemes share a strong supposition that the 'problem' of global environmental governance can largely be reduced to, and resolved by, playing around with the design of global environmental organizations. It is the fundamental flaws of this premise, and the dangers of taking it too seriously, that this chapter will focus on.

2   A Dangerous Confusion: 'Institutions' and 'Organizations'

Although the WEO/GEO literature routinely refers to its enterprise in terms of 'institutions', it tends to use the term as if it were the plural of 'organization'. The distinction, of course, is *not* merely semantic, is well established in the literature, and is critical to this context (Young 1994). Institutions, as von Moltke reminds us, are 'social conventions or "rules of the game", in the sense that marriage is an institution, or property, markets, research, transparency or participation' (von Moltke 2001b, 11). Therefore, institutions need not necessarily have a physical existence. Organizations, on the other hand, are much more circumscribed. According to Oran Young, they are 'material entities, possessing physical locations (or seats), offices, personnel, equipment, and budgets' (Young 1989, 32). The WEO/GEO discourse is clearly preoccupied with organizations and often ignores fundamental questions about why environmental degradation happens, or why global cooperation founders, or even why global environmental governance is a good idea (Newell 2001).

This confusion has the effect of trivializing global environmental governance. To place the spotlight on organizational tinkering and label it 'institution building' is to imply that the 'institutional will'—in terms of societal conventions and 'rules of the game'—for global environmental cooperation already exists and all that remains is to set up an appropriate organizational framework (von Moltke 2001a); that global cooperation is a function of inappropriately designed organizations, rather than a reflection of a fundamental absence of willingness on the part of states (Juma 2000); that the lack of implementation of international regimes stems from dispersed secretariats, rather than the failure of these very same regimes 'to target those actors that create the problems that regime arrangements set out to address' (Newell 2001, 40); and that improved global environmental governance is a puzzle of administrative efficiency, rather than a challenge of global justice (Agarwal et al. 1999). None of the institutional challenges identified here are likely to be resolved by merely rearranging the organization of chairs on our planetary Titanic. Unless we somehow address the core institutional questions first, any new organization will fall prey to the exact same pathologies that confront existing arrangements.

The focus on organizational minutiae is dangerous precisely because it distracts from the more real and immediate institutional challenges to global environmental governance. Two such challenges are of particular importance; both are treated only peripherally by GEO/WEO proponents, if at all.

The first relates to near demise of the much-celebrated Rio compact on sustainable development—the supposed understanding between the developing countries of the South and their more industrialized counterparts from the North that environment and development will be dealt with as an integrated complex of concerns within the context of current and future social justice and equity (Najam 2002b). The compact, to whatever extent it did exist, was always understood to be an expression of desire rather than reality—what Tariq Banuri (2001) has called 'a triumph of hope over experience'.[2] The hope, obviously misplaced, was that once the compact would become real, both North and South would somehow learn not simply to accept it but to operationalize it; but that was not to be (Banuri 1992, Najam 1995, Sandbrook 1997, Agarwal et al. 1999). In fact, the optimism was shed rather quickly—the North soon became wary of the 'fuzziness' of sustainable development while the South began to fear that the supposed 'definitional' problems with the concept were being used as an excuse for maintaining the status quo (Najam 2003b). As the World

---

[2]   The phrase is based on an original attributed to Mark Twain.

Summit on Sustainable Development (WSSD) came along, the concept was already much-bruised, much-diluted, and ready to be buried in Johannesburg. In the run-up to WSSD a number of voices, including the literature already cited, began calling for new governance arrangements for the environment. Some political momentum was added to these efforts as both UNEP and the World Bank showed mild interest in talking over the cause of organizational re-jigging. The discussions that ensued in the run-up to WSSD did little to spruce up the reality of the failed Rio bargain, and the attempt to get a new organizational architecture, or at least a spruced up organizational architecture, at WSSD proved futile.

The implication of that failed debate for the future of global environmental governance is significant. Even more important are the lessons we might learn from that halted debate. To whatever extent the concept of sustainable development had originally embodied the semblance of an *institutional* bargain on how environmental issues should be contextualized globally, that bargain is now largely defunct—and so is the very tentative and always nebulous accord that might once have existed on why global environmental governance may be a good thing, for whom, and on what terms. It is not a surprise, then, that the immediate reaction of many in the South is to shirk at the first mention of a GEO or a WEO; or that the addition of development-related flourishes to these proposals fail to woo the South and are either rejected to ignored (Agarwal et al. 1999, Newell 2001). Frankly, the sometimes glib and always lofty goals of finding 'thoughtful ways to manage our ecological interdependence' (Esty 2000, 14) or of '[elevating] environmental policies on the agenda of governments, international organizations, and private actors' (Biermann 2000, 29) or even of 'equitably and effectively [managing] planet Earth' (Biermann 2002, 29) are no longer credible, or necessarily appealing, to those who have lived through more than a decade of broken global promises on sustainable development. In essence, the very basis of global environmental cooperation—and thereby governance—that might have seemed to exist a decade ago, is under threat today. As Agarwal and colleagues (1999, 37) point out, no effective governance is possible under the prevailing conditions of deep distrust. Hence, organizational rearrangements might distract from deeper problems but are unlikely to solve them.

The second critical challenge to the cause of improved global environmental governance pertains not to the exclusion of the concerns of

Southern governments from the emerging 'New Global Environmental Order' but to the meaningful inclusion of civil society concerns—especially those of the South (Agarwal et al. 1999). This is important because the very nature of the environmental problematique is different from many other international concerns (such as defence and security) in that a greater proportion of key environmental decisions lie beyond the direct ability or authority of states. This underscores the need for a society-centric view of global environmental governance that includes state organs but goes beyond them (Banuri and Spanger-Siegfried 2000). This, of course, stands in contradiction to the predominantly state-centred view of global governance in the organizationally inclined literature. This is not to suggest that interstate organizations are unimportant. Far from it, they will have to be an integral—probably a central—component of improved global environmental governance. From an institutional perspective, however, the quality of such governance will be determined by how interstate organizations connect with emerging global public policy networks, of which civil society organizations are a key part (Reinicke 1998). In ignoring, downplaying, or at the very least distracting from the centrality of such integration the organizational debate fails to rise to the challenge of what could have been a very timely discourse on meaningful 'institutional' reform.

Having said the above, there are other streams of scholarship on global governance that do recognize the key challenge as the creation of institutions that can integrate the multitude of voices that now feel alienated from the official chatter on global environmental issues. For example, those who talk in terms of global public policy networks, see better governance emanating not just from decisions taken at centralized interstate organizations or via coordinated legal frameworks but also through networks of dispersed decision points spread out globally, across all sectors—state, market and civil (Reinicke and Deng 2000). This leads one to a very different set of organizational questions. The emphasis would shift from a search for better management as measured by administrative efficiency to better networking as gauged by broad-based legitimacy (Banuri and Spanger-Siegfried 2000).

The centrality that has been assumed by the organizational debate within the global environmental discourse translates to a distraction from these other pressing issues. It is not only that new organizational manoeuvring is likely to be insufficient to revive the spirit of the Rio compact or to integrate with civil society networks, it is also that any new organizational arrangement is likely to remain as stymied as the current arrangement until these other issues of global environmental governance are tackled *first*.

3 'New Lamps for Old'

Ever since *Aladdin and the Magic Lamp*,[3] those of us who come from what used to be called the Orient and is now a part of the larger global 'South' have learnt to be wary of anyone offering 'new lamps for old!'. Therefore, when someone offers to replace existing organizational arrangements with a 'new and improved' architecture, one instinctively asks: 'What is it that is so bad about the old or so different about the new?'. In the case of global environmental organizations the answer is, 'Not much!'.

Proponents of organizational rearrangement invariably begin with the standard scare tactics—global ecological systems are under growing threat. While this assessment is correct in and of itself, the jump between acknowledging the ecological crisis and pointing to organizational inefficiency as the culprit is a rather wide one. Beyond assertion, there is no attempt to establish causality, or even correlation, between the continuing ecological crisis and the nature of the existing organizational arrangements. Two questions, it seems, need to be asked. First, would things have been worse had the existing system *not* been in place? As we will argue later, the answer to this is certainly that yes, they would have been worse. Second, could things be better under an alternative system?

Proponents of large-scale organizational rearrangement obviously believe that things would, in fact, be improved if we rearranged the organizational architecture. They accuse the existing arrangements of a coordination deficit, deficient authority and insufficient legitimacy and

---

3 In this famous children's tale—one of the original Arabian Nights—the villain is able to steal the magic lamp from Aladdin's unsuspecting wife by making her an offer she cannot refuse: promising to replace old lamps for new.

promise that setting up a new organization would streamline organizational coordination, accelerate financial and technology transfers, and improve the implementation and development of international environmental law (Esty 1996, Biermann 2000). What is not made clear, however, is *why* the pathologies that inflict the existing arrangements would simply not be transferred to any new arrangement? If coordination is the real roadblock to better environmental performance, then why should one believe that a new organization could achieve it better than UNEP? After all, UNEP's very *raison d'être* has been to coordinate and catalyze. Why should one assume that rich nations that have been so stingy in meeting their global fiscal responsibilities in the past—in environmental as well as other arenas—will suddenly turn generous for a new organization? If fragmentation is what makes the current arrangements unwieldy, could that not be addressed within the framework of Section 38.22 (h) of Agenda 21, which called for the co-location of various treaty secretariats under the UNEP umbrella? What in the new system would make Northern governments—who have consistently reneged on their international commitments regarding financial and technology transfer—suddenly reverse this trend? In short, the most interesting questions are never asked, and certainly not answered.

The problems that these proposals seek to solve through reorganization are not organizational problems at all. If UNEP has been denied authority and resources, it is because the nation states wish to deny it (and any successor super-organization) authority and resources. They have certainly never demonstrated the willingness to provide UNEP with the resources that would be required to do what they claim it ought to do. The 'coordination deficit' is not something that crept in. It was something that was painstakingly designed into the system because the countries that are most responsible for the global ecological crisis have never demonstrated the intention of owning that responsibility and because intense turf battles between UN agencies forced an unmanageable coordination mandate upon UNEP (Gosovic 1992). The 'coordination deficit' is indeed real, but it is not organizational—it is institutional and is unlikely to go away through cosmetic architectural renovations. With due apology for sounding cynical, the point to be made is that the crisis at hand is not about organizational minutiae but about the now glaring lack of willingness for global environmental cooperation, particularly but not solely from the United States.

The problems that the proponents of organizational rearrangement identify are, for most part, real problems. The goals they identify for the rearranged system are laudable goals. One has no qualms with either. The issue is how the dots are connected or, in this case, not connected. The proposals inspire no confidence that the problems confronted by the current setup will not simply transfer to a new setup, or that new arrangements would be any more likely to succeed where the current arrangements have failed. Rather, this seems to be one more incidence of 'hope triumphing over experience'.

4 Viva la UNEP

Although not always intentional, the immediate casualty of the misdiagnosis on the part of proponents of a world environment organization is the reputation of the United Nations Environment Programme. Even though some (including some contributors of this volume) view UNEP as the central core of the ultimate superstructure for global environmental governance—and some within UNEP may well find this notion appealing—the fact of the matter is that, implicitly or explicitly, UNEP is portrayed as being at the root of the 'problem'. After all, when the existing organizational structure is accused of being inefficient, ineffective and insufficiently legitimate, then UNEP—which is the centrepiece of that structure—must also stand accused, even if indirectly. Indeed, one should concede that like any other UN agency, UNEP has much that can be improved. However, the stings—implied or explicit—showered on it either ignore or underplay its very significant achievements.

The tragedy is not just that such proposals are based on the assumption that the much-trumpeted 'weakness' of UNEP lies at the heart of the crisis of global environmental governance. Nor is it just that even the critics of such schemes nearly never question this assumption. The real tragedy is that UNEP's own leadership seems to have bought into this assumption. The rampage of exaggerated external criticism and unwarranted self-doubt cannot bode well for UNEP or for the morale of its staff. Indeed, this chapter argues that while UNEP is certainly not the 'perfect' agency, and while there is much that can and should be improved, it is *not* the weakling or underachiever that it is generally portrayed as.

Arguably, it has performed relatively well in comparison to other agencies of the UN family in terms of both performance and legitimacy, and it has every right to stand proud of its achievements that, by the way, came despite all the limitations that its critics are so fond of enumerating.

Like much of what is being proposed in the current round of debate, UNEP was originally conceived as the 'environmental conscience of the UN system' and was charged to act as the 'focal point for environmental action within the United Nations system'.[4] In defining this mandate of coordination, it was thrust with 'perhaps one of the most difficult jobs in the entire UN system' (Sandbrook 1983, 388). It has been hinted at that UNEP may have been designed for failure—or at least for something less than success (von Moltke 1996). As McCormick (1995, 152) points out 'it had severe obstacles placed in its path from the outset. It had too little money, too few staff, and too much to do. It had the thankless task of coordinating the work of other UN agencies against a background of inter-agency jealousy and suspicion, and national governments were unwilling to grant UNEP significant powers'. Given the sprawling and bickering nature of the UN machinery, its own lack of executive status, and the dismal resources at its command, 'UNEP could no more be expected to "coordinate" the system-wide activities of the UN than could a medieval monarch "coordinate" his feudal barons' (Imber 1994, 83). It should, therefore, be no surprise that UNEP has not been able to fulfil, what Conca (1996, 108) has called, 'its hopeless mandate as system-wide coordinator on environmental matters'.

Yet, while there is agreement that UNEP has not been allowed to fulfil its coordination mandate, it is also argued that it 'can be credited with having achieved more than it was in reality empowered to do' (McCormick 1995, 153). Those who have studied it at depth agree that it is 'generally well-regarded' (Imber 1993: 56), 'relatively effective' (Conca 1996, 112), and given its meagre resources and authority it 'has been a remarkable success' (von Moltke 1996, 58). While this is not the place to evaluate UNEP's

---

4   United Nations General Assembly Resolution 2997 (XXVII), adopted December 1972.

achievements, let us list a sampling of reasons why it should be considered a successful international organization.

Making the Environment a 'Global' Concern. The single most important and much unappreciated achievement of UNEP is its role in converting the environment into a 'global' concern. It is easy to forget the hostility with which the developing countries had greeted the Stockholm Conference of 1972 and the subsequent establishment of UNEP (see Founex Report 1972). The placement of UNEP in Nairobi was not just a 'symbolic' act. It was a strategic necessity without which the developing countries might never have accepted an environmental organ to be created (Rowland 1973). The fact that this became the first United Nations organ to be based anywhere in the developing world galvanized the South both in the process of getting it to locate in Nairobi and in its early and most difficult years from the 1970s into the mid-1980s. This period also marked the height of Southern solidarity and the movement for a 'New International Economic Order' (NIEO). During this period the symbolism of UNEP being set up in Nairobi was of significant importance to the developing countries. This resulted in their very visibly and disproportionately supporting an organization that they had originally resented. Although they stood with UNEP largely out of a sense of Southern solidarity, the developing countries began buying into parts of the environmental agenda and, more importantly, demanding that the agenda be modified to incorporate their realities. Indeed, the call to set up the World Commission on Environment *and* Development (WCED) came out of a discussion at the UNEP Governing Council. While WCED might have come up with the term 'sustainable development', the stage for it had already been set by UNEP and its Governing Council at its tenth anniversary meeting in 1982.

Advancing the Global Environmental Agenda. Those who gathered at Stockholm in 1972 could scarcely have imagined the global environmental agenda becoming as advanced and as prominent in international affairs as it is today. UNEP played a significant part in this transformation (Caldwell 1996). Through its various activities, and especially training programmes, it helped create an environmental constituency within and outside governments that has been at the forefront of moving this agenda forward. It played a pivotal role in putting desertification, ozone depletion, and organic pollutants on the global agenda (Downie and Levy 2000). Even for issues like climate change, biological diversity and deforestation, UNEP's contribution has been more important than it is often given credit for.

International Environmental Law. International environmental law has probably been the single fastest growing sub-field of international law, and the United Nations Environment Programme has been amongst the most active and productive UN agencies in terms of advancing international law. This is not an idle statement. Apart from the agenda-setting role it played on issues such as desertification, biological diversity and climate change, it has been the principal negotiation-manager for complex global regimes on ozone depletion, trade in endangered species, trade in hazardous wastes, persistent organic pollutants, regional seas, et cetera. For an organization as young and as resource-strapped as UNEP, this is a remarkable achievement indeed. Importantly, UNEP-managed treaty negotiations— such as those on ozone-depleting substances and more recently on persistent organic pollutants—are generally regarded to have been amongst the most efficient and successful global environmental negotiations (Benedick 1991, Tolba 1998).

Legitimacy. By routinely suggesting that a new organizational architecture would lend legitimacy to global environmental governance, the proponents of a world or global environment organization seem to imply that UNEP has less than sufficient legitimacy as an international organization. If they were not earnest, it would be funny that some proponents of a super-organization wish to scrap UNEP and replace it with something that might look more like the World Trade Organization (WTO). Massive public demonstrations from Seattle to Prague and feelings of distrust and apprehension is what comes to mind when one thinks of the WTO or the World Bank (another organization that is sometimes talked about as the model to follow). UNEP, on the other hand, does not have to place barriers or bring out riot police at its annual meetings and has a tradition of good relations with civil society. Indeed, in terms of general public legitimacy and honest efforts to involve civil society in its orbit, UNEP has fared much better than most international organizations even though there remains room for improvement (Banuri and Spanger-Siegfried 2000).

In sum, while UNEP has its share of problems they relate not to its mandate as much as to the resources that have been provided to it. The fact that some of its critics have never forgiven it for being located in a developing country does not help either. It is unfortunate that its leadership has sometimes been defensive about both its achievements and its potential, instead of building upon its rather rich legacy of performance. It is by no means a perfect organization, but it has been a rather good one. It would be sad if in our zeal for organizational rearrangement, we made the allegedly perfect the enemy of the demonstrably good.

5   Towards Better Global Environmental Governance

It should be obvious that this author is not persuaded by the need for an environmental super-organization. However, an argument against new organizational superstructures should not be confused with an argument for organizational inertia. All organizations should strive for improvement, and global environmental organizations—including the United Nations Environment Programme—are no exception. There are a number of elements within the various proposals that do make sense—not as arguments for organizational rehaul, but as elements of an agenda to improve the existing organizational setup. Moreover, change that happens within the existing system is likely to be substantively less disruptive and politically more feasible. This final section highlights five key elements of a potential agenda for organizational improvement that can be pursued within the confines of the existing structures and would mark a beginning in addressing the larger institutional challenges of global environmental governance as discussed earlier.[5] It should be noted, however, that for all the reasons already discussed, some of these ideas are not going to be *easy* to implement. Yet, to the extent they can be implemented, they are likely to

---

[5]   Some of these ideas are informed by the author's participation at the 'Expert Consultation on International Environmental Governance' organized by UNEP in Cambridge, UK, on 28-29 May 2001. A report of the meeting is available at http://www.unep.org/IEG/docs/.

be *easier* to implement within UNEP's existing structure than within a new supra-organization of the GEO/WEO variety.

Enabling UNEP to Fulfil its Mandate

There is no need to change UNEP's mandate. There is, however, an urgent need to provide it with the resources, staff and authority it needs to fulfil its mandate. UNEP's shareholders—i.e. the member states—need to invest in UNEP in proportion to the responsibilities that they demand of it. One step in this direction might be to convert UNEP into a specialized agency (as opposed to a 'Programme') with the concomitant ability to raise and decide its own budget. Greater autonomy may not, in itself, be sufficient to translate to greater resources but it could allow the organization to be more innovative and even assertive in its recourse mobilization strategies. However, given the political wrangling this would require, the UN General Assembly might consider maintaining UNEP's 'Programme' status but providing it with greater autonomy in budgetary matters to ensure a sufficient and consistent resource base.

Indeed, UNEP was originally modelled around the United Nations Development Programme (UNDP) and should aspire to fulfil that original intent. While this would obviously require the UN Secretary-General and members states to give UNEP the budgetary and operational prominence that it has so often been promised, it would also require more assertive leadership from UNEP so that it gets the respect it deserves. One step could be to invest in making its flagship *Global Environmental Outlook* reports a more prominent product, with the goal being to make it an environmental equivalent to the World Bank's *World Development Reports*, or UNDP's *Human Development Reports* in terms of influence and recognition.

Realizing Sustainable Development

Over the years, many have become quite fond of arguing that the problem with sustainable development is that it is very difficult to define. While defining it in precise terms is certainly not easy, it is also not entirely necessary. The real problem with sustainable development is that the governments of the world lack the commitment to realizing it. The main culprits in this regard are governments in the North that have consistently reneged on their financial commitments. However, the governments of the South are also to blame for viewing sustainable development simply as an excuse to continue with development as usual without any regard to its environmental consequences (Najam 2002a).

From an institutional perspective, realizing sustainable development would imply streamlining mechanisms for financing sustainable development, and monitoring and validating progress. Because of problems of transparency and performance, many developing countries consider the Global Environment Facility (GEF) to lack legitimacy (Agarwal et al. 1999). Other funding mechanisms are even more strapped for cash. Two major events in 2002—the International Conference on Financing for Development in Mexico and the World Summit on Sustainable Development in South Africa—provided an ideal opportunity to reconsider the operation of GEF, broaden the scope of activities that it can finance, replenish it to higher levels, and possibly place its management more firmly within UNEP, which enjoys more credibility with developing countries and routinely deals with issues of environment and sustainable development as its primary focus. Both these opportunities were largely wasted, although some small headway was made in terms of GEF replenishment. However, what is needed are bolder structural steps. For example, the existing trilateral management structure involving UNEP, UNDP and the World Bank can, in fact, be maintained while UNEP is given the role of the 'lead' agency in its actual management. Doing so would also go a long way in allowing UNEP to better fulfil its existing mandate.

Managing the Proliferation of MEAs

Over the last decade, the great increase in negotiations pertaining to the new or existing multilateral environmental agreements (MEAs) has caused a serious problem of 'MEA proliferation' and attendant pathologies of 'negotiation fatigue', particularly amongst developing country delegates (Najam 2000, 2002b). This has placed an immense burden on most developing countries, which simply do not have the resources to keep up with the frantic pace of increasingly complex negotiations. Moreover, the frenzy to complete negotiations as quickly as possible has left behind a legacy of less-than-perfect agreements or resulted in too little attention being paid to questions of implementation (Najam and Sagar 1998, Najam et al. 2003, Najam 2004).

It makes sense to pause and consider how these various MEAs fit together. A certain clustering of independently negotiated treaties has begun to emerge organically as part of the evolution of international environmental law and it appears timely to convert this into a deliberate scheme (Najam 2002). Von Moltke, in particular, has outlined a useful list of possibilities for clustering MEAs (von Moltke 2001b and this volume). A co-location of MEA secretariats—which is also suggested by some proponents of a world environment organization (Biermann, this volume)— seems an equally pragmatic idea even though it is likely that some governments and secretariat staff might resist it. Yet, it is an idea worth pursuing because it could provide efficiency gains, increase cross-treaty communication, and deter from MEA fiefdoms. Overlapping or joint meetings of related MEAs, possibly in permanent locations, would serve to ease the pressures on participating delegates and encourage more continuity in representation.

The United Nations Environment Programme, with its good record of managing MEAs—both in terms of overseeing complex negotiations and of hosting MEA secretariats—again emerges as the best-suited candidate for this job. However, this would *not* require a new super-organization. Nor would it require a major legal restructuring of UNEP's mandate. The task was already awarded to UNEP a decade ago by *Agenda 21* (Section 38.22 [h]), which called upon UNEP to 'concentrate' on (amongst other things), the 'further development of international environmental law, in particular conventions and guidelines, promotion of its implementation, and coordinating functions arising from an increasing number of international legal agreements, *inter alia*, the functioning of the secretariats of the Conventions ... including possible co-location of secretariats established in the future'.

Coordination, Yes—Centralization, No

Echoing UNEP's original charter, *Agenda 21* (Section 38.23) had also defined UNEP as the 'principal body within the United Nations system in the field of environment'. However, for good reason, neither had seen it as the *only* UN body with relevance to the environment. Centralization makes little conceptual sense for issues related to the environment, and even less so for sustainable development. The fabric of environmental concerns, and even more with regard to sustainable development concerns, is a multivariate web of interlinked issues that do not have a clear 'centre' and are unlikely to respond to centralized policy-making. A reading of *Agenda 21*, or of the report of the World Commission on Environment and Development before it, would make it quite clear that if there is any body that has the authority to centrally devise all of sustainable development policy—or to even coordinate all of sustainable development policy—that body would be the United Nations Organization as a whole rather than any sub-component of it. Could one imagine creating a central entity that is responsible for all issues related to sustainable development ranging from biological diversity to international debt, from climate change to education, from poverty alleviation to pollution? Should one imagine such a centralized entity, even if one could?

Given the fundamentally interlinked and cross-sectoral nature of these issues, UNEP's original mandate as a catalyst and coordinator was, in retrospect, quite well-conceived. However, as already noted, UNEP has been less than successful in realizing its coordination mandate. At the same time, the coordination mandate is now spread out around the system—in addition to the Commission for Sustainable Development, the recently created Environmental Management Group and the Global Ministerial Environment Forum (GMEF) both seem to have some elements of the coordination function in their mandates. This dilution of UNEP's coordination responsibility may not be a bad thing. Not only is coordination a thankless job but also, as Mark Imber reminds us, 'the primary responsibility for coordination rests with governments' (Imber 1993, 66). The agency heads that constitute the Environmental Management Group and the senior government delegates that make up the CSD and GMEF respectively seem far better positioned for UN-wide coordination than UNEP's secretariat staff could ever be expected to. Having multiple forums for coordination may also not be bad—there is enough cross-participation within these groups to keep duplication or contradiction manageable, while multiple forums could actually have the effect of reinforcing each other on the need for coordination.

'Civilizing' Global Environmental Governance

Providing the space and opportunity for meaningful participation of civil society networks in global environmental governance may well be the most important challenge from the institutional, as well as the organizational standpoint (Banuri and Spanger-Siegfried 2000). Within the realm of global public policy, the environment is an issue where civil society has been particularly active and influential (Gordenker and Weiss 1995, Najam 1999). However, there is a growing sense that international organizations are becoming increasingly introverted. Especially in the aftermath of recurrent civil protests arising from a deeply felt distrust of globalization—and of international organizations as the agents thereof—both UNEP and the Commission for Sustainable Development need to invest more attention to linking with civil society. In a recent report, Banuri and Spanger-Siegfried (2000) lay out a detailed set of recommendations for establishing deeper linkages with civil society actors, particularly global public policy networks, for leveraging the opportunities for policy innovation and cross-sectoral synergies that this would offer.

We also need to begin viewing civil society not just as stakeholders in, but also as motors of global environmental governance. Following the tradition of human rights regimes, civil society networks could potentially become the real drivers of MEA implementation. Indeed, for political as well as logistic reasons, they may be more likely to play that role than governments or intergovernmental agencies. There is growing evidence already that civil society actors have emerged as the most vibrant 'civic entrepreneurs' for sustainable development (Banuri and Najam 2002). The need is to nurture these civic entrepreneurs and to provide them with forums for congregation and dialogue both within themselves and with the interstate and state policy decision-makers.

6   Conclusion

In conclusion, the argument of this chapter is that not only do we not need a new international environmental organization, but also the discussions on this subject tend to distract from actual reforms in the existing organizations that are indeed required. The chapter outlines five elements of such an agenda for organizational reform, but recognizes that these elements must be embedded in the larger challenge of institutional reform. In practical terms, this means that the key change has to come not in the structural details of existing or new organizations, but in the support and political will that national governments are willing to invest in these organizations.

References

Agarwal, A., S. Narain and A. Sharma (1999), *Green Politics*, Centre for Science and Environment, New Delhi.

Banuri, T. (1992), 'Noah's Ark or Jesus's Cross?', Working Paper WP/UNCED/1992/1, Sustainable Development Policy Institute, Islamabad.

Banuri, T. (2001) 'Envisioning Sustainable Development', Stockholm Environmental Institute, Boston [unpublished note].

Banuri, T. and A. Najam (2002), *Civic Entrepreneurship: Civil Society Perspectives on Sustainable Development*, Gandhara Academy Press, Islamabad.

Banuri, T. and E. Spanger-Siegfried (2000), *UNEP and Civil Society: Recommendations for a Coherent Framework of Engagement*, Stockholm Environmental Institute, Boston.

Benedick, R. (1991), *Ozone Diplomacy: New Directions in Safeguarding the Planet*, Harvard University Press, Cambridge (Mass.).

Biermann, F. (2000), 'The Case for a World Environmental Organization', *Environment*, vol. 42 (9), 22–31.

Biermann, F. (2002), 'Strengthening Green Global Governance in a Disparate World Society: Would a World Environment Organization Benefit the South?', *International Environmental Agreements: Politics, Law and Economics*, vol. 2, 297–315.

Biermann, F. (2005), 'The Rationale for a World Environment Organization', in F. Biermann and S. Bauer (eds), *A World Environment Organization. Solution or Threat for Effective International Environmental Governance?*, Ashgate, Aldershot, 117–143.

Caldwell, L.K. (1996), *International Environmental Policy: Emergence and Dimensions* (3rd Edition), Duke University Press, Durham.

Charnovitz, S. (1995), 'Improving Environmental and Trade Governance', *International Environmental Affairs*, vol. 7 (1), 59–91.

Conca, K. (1996), 'Greening the UN: Environmental Organizations and the UN System', in T.G. Weiss and L. Gordenker (eds), *NGOs, the UN and Global Governance*, Lynne Rienner, Boulder, 103–119.

Downie D.L. and M.A. Levy (2000), 'The UN Environment Programme at a Turning Point: Options for Change', in P.S. Chasek (ed.), *The Global Environment in the Twenty-First Century: Prospects for International Cooperation*, United Nations University Press, Tokyo, 355–77.

Esty, D.C. (1994), 'The Case for a Global Environmental Organization', in P.B. Kenen (ed.), *Managing the World Economy: Fifty Years After Bretton Woods*, Institute for International Economics, Washington, DC, 287–309.

Esty, D.C. (1996), 'Stepping Up to the Global Environmental Challenge', *Fordham Environmental Law Journal*, vol. 8 (1), 103–13.

Esty, D.C. (2000), 'The Value of Creating a Global Environmental Organization', *Environment Matters: Annual Review* (July 1999–June 2000), World Bank, Washington DC, 13–15.

Founex Report (1972), *Development and Environment*, Report and Working Papers of Experts Convened by the Secretary General of the United Nations Conference on the Human Environment held at Founex, Switzerland, 4–12 June 1971, Mouton, Paris.

Gordenker, L. and T.G. Weiss (1995), 'NGO participation in International Policy Processes', *Third World Quarterly*, vol. 16 (3), 543–55.

Gosovic, B. (1992), *The Quest for World Environmental Cooperation: The Case of the UN Global Environment Monitoring System*, Routledge, London.

Imber, M. (1993), 'Too Many Cooks? The Post-Rio Reform of the United Nations', *International Affairs*, vol. 69 (1), 55–70.

Imber, M. (1994), *Environment, Security and UN Reform*, St. Martin's Press, London.

Juma, C. (2000), 'The Perils of Centralizing Global Environmental Governance', *Environment Matters: Annual Review* (July 1999–June 2000), World Bank, Washington, DC, 13–15.

McCormick, J. (1995), *The Global Environmental Movement*, John Wiley and Sons, New York.

Najam, A. (1995), 'An Environmental Negotiation Strategy for the South', *International Environmental Affairs*, vol. 7 (3), 249–87.

Najam, A. (1999), 'Citizen Organizations as Policy Entrepreneurs', in D. Lewis (ed.), *International Perspectives on Voluntary Action: Reshaping the Third Sector*, Earthscan, London, 142–81.

Najam, A. (2000), 'The Case for a Law of the Atmosphere', *Atmospheric Environment*, vol. 34 (23), 4047–49.

Najam, A. (2002a), 'Financing Sustainable Development: Crises of Legitimacy', *Progress in Development Studies*, vol. 2 (2), 153–60.

Najam, A. (2002b), 'The Unraveling of the Rio Bargain', *Politics and the Life Sciences*, vol. 21 (2), 46–50.

Najam, A. (2003a), 'The Case Against a New International Environmental Organization', *Global Governance,* vol. 9 (3), 367–84.

Najam, A. (2003b), 'Innocence Lost: Developing Countries in Post-Rio Environmental Negotiations', in Sustainable Development Policy Institute (ed.), *Sustainable Development and Southern Realities: Past and Future in South Asia,* City Press, Karachi, 173–84.

Najam, A. (2004), 'Developing Countries and the Desertification Convention: Portrait of a Tortured Relationship', accepted for publication *Global Environmental Politics.*

Najam, A. and A. Sagar (1998), 'Avoiding a COP-out: Moving Towards Systematic Decision-Making Under the Climate Convention', *Climatic Change,* vol. 39 (4), iii–ix.

Najam, A., S. Huq and Y. Sokona (2003), 'Climate Negotiations Beyond Kyoto: Developing Country Concerns and Interests', *Climate Policy,* vol. 3 (4), 221–31.

Newell, P. (2001), 'New Environmental Architectures and the Search for Effectiveness', *Global Environmental Politics,* vol. 1 (1), 35–44.

Reinicke, W.H. (1998), *Global Public Policy: Governing without Government?,* Brookings Institution Press, Washington, DC.

Reinicke, W.H. and F.M. Deng (2000), *Critical Choices: The United Nations, Networks and the Future of Global Governance,* International Development Research Center, Ottawa.

Rowland, W. (1973), *The Plot to Save the World,* Clarke, Irwin and Co., Toronto.

Runge, F.C., F. Ortalo-Magne and P. van de Kamp (1994), *Freer Trade, Protected Environment: Balancing Trade Liberalization and Environmental Interests,* Council on Foreign Relations Press, New York.

Sandbrook, R. (1983), 'The UK's Overseas Environmental Policy', in *The Conservation and Development Programme for the UK: A Response to the World Conservation Strategy,* Kogan Page, London.

Sandbrook, R. (1997), 'UNGASS has Run Out of Steam', *International Affairs,* vol. 73 (4), 641–54.

Tolba, M.K. (1998), *Global Environmental Diplomacy: Negotiating Environmental Agreements for the World, 1973–1992,* MIT Press, Cambridge (Mass.).

von Moltke, K. (1996), 'Why UNEP Matters', in *Green Globe Yearbook of International Co-operation on Environment and Development,* Oxford University Press, Oxford, 55–64.

von Moltke, K. (2001a), 'The Organization of the Impossible', *Global Environmental Politics,* vol. 1 (1), 23–28.

von Moltke, K. (2001b), *Whither MEAs? The Role of International Environmental Management in the Trade and Environment Agenda,* International Institute for Sustainable Development, Winnipeg (Canada).

von Moltke, K. (2005), 'Clustering International Environmental Agreements as an Alternative to a World Environment Organization', in F. Biermann and S. Bauer (eds), *A World Environment Organization. Solution or Threat for Effective International Environmental Governance?*, Ashgate, Aldershot, 175–204.

Whalley J. and B. Zissimos (2001), 'What Could a World Environmental Organization Do?' *Global Environmental Politics*, vol. 1 (1), 29–34.

Young, O.R. (1997), 'Global Governance: Toward a Theory of Decentralized World Order', in O. Young (ed.), *Global Governance: Drawing Insights from the Environmental Experience*, MIT Press, Cambridge (Mass.), 273–99.

Young, O.R. (1989), *International Cooperation: Building Regimes for Natural Resources and the Environment*, Cornell University Press, Ithaca, NY.

Young, O.R. (1994), *International Governance: Protecting the Environment in a Stateless Society*, Cornell University Press, Ithaca, NY.

Chapter 10

Conclusion

Frank Biermann and Steffen Bauer

As evidenced from the contributions to this volume, no end is in sight to the debate on a world environment organization that has begun more than thirty years ago. Given persistent disagreements, the discussion is likely to continue, both in diplomatic negotiations between countries that support a new organization and those that oppose it, and in academic circles that seek to provide input into and comment on these deliberations.

At the same time, this volume shows that points of agreement are emerging, perhaps even a convergence of views towards some 'middle ground'. Most observers now support strengthening the United Nations Environment Programme, with an enlarged mandate and a more predictable financial basis. Also, most observers agree that major revolutionary change is neither feasible nor desirable: the abolishment or merger of major international agencies, the creation of new big bureaucracies, or the setting-up of international bodies with strong enforcement powers—such reform visions, which are still found in the literature, are not likely to muster much support today. The debate on a world environment organization has contributed to both these points of convergence: it has pushed the insufficient mandate of and lack of governmental support for UNEP on the agenda of international deliberations while shaping the debate in a way that has filtered the more feasible reform options from the more radical grand designs.

The contributions to this volume stand witness to this evolution. The complexity and richness of the debate is illustrated by Joyeeta Gupta's chapter, in which she lists nine major options for a design of global environmental governance. This list includes a world environment organization, although Gupta remains sceptical regarding the alleged improvements this would entail for the South. Similarly, the contribution of Lorraine Elliot outlines in broad strokes the current crisis of the global environmental governance system and puts the debate into the context of the post-Rio process. She recalls the historical compromise that allowed Northern and Southern interests to be fused rhetorically under the formula of sustainable development—rather than environment *or* development—but failed to deliver an effective governance structure that could live up to it.

Regarding outspoken proponents of a WEO, the contribution of Steve Charnovitz, who has participated in the debate for 15 years, provides a highly insightful review on how the discourse has influenced his perception of the need for a world environment organization. Charnovitz remains faithful to his original idea of a new agency (he prefers to call it a 'global environment organization'), which he now presents in a refined manner that takes recent criticism into account. The same holds for Biermann's chapter, which presents an evolved version of his original argument. He develops further his proposal to upgrade UNEP to a world environment organization, which in his view would serve both North and South better than the current status quo. John Kirton provides a valuable contribution on the particular perspective of industrialized countries, which bear the main responsibility for many current global environmental problems. He focuses less on the Southern perspective and the prevalent fears among governments in developing countries of Northern 'eco-colonialism' through global environmental regulation. Consequently, Kirton locates his model of a world environment organization outside the traditional UN system—not unlike George Kennan's original call for an international environmental agency. An interesting subject for further research would be to examine how in Kirton's design the 'Group of 8' could move forward in creating an organization that would eventually ensure universal participation by integrating also developing countries.

An explicit critique of the idea of a world environment organization is offered by Sebastian Oberthür and Thomas Gehring, two scholars of international relations who have participated in the debate on organizational reform for half a decade. They argue that fragmentation of policies and regimes is less a problem than an asset. Decentralized approaches will, in their view, allow governments to develop better solutions exactly designed for each environmental problem at hand. On the other hand, they do not elaborate much on why environmental policy differs from other policy areas, such as labour, health or trade, all of which function through centralized negotiations and regime-creation processes. Furthermore, proponents of a new organization often do not deny the importance of decentralized approaches, but rather see them, together with international regimes, as two equally important elements of an effective governance architecture.

A different argument against a world environment organization is brought forward by Adil Najam. According to Najam, the very quest for a world environment organization results from a misdiagnosis of the current governance crisis and neglects key underlying issues that would need to be addressed first. Most of his critique is directed against grand reform designs, such as proposals of new agencies that would have sanctioning powers vis-à-vis member states, as well as 'organizational tinkering' at large. In the end, however, Najam, too, supports a considerable strengthening of UNEP and even suggests 'to convert UNEP into a specialized agency (as opposed to a "Programme") with the concomitant ability to raise and decide its own budget' (this volume, p. 248)—this is precisely what many supporters of a world environment organization have in mind. Najam's approach is thus, despite his rhetoric, not that distinct from moderate supporters of a world environment organization, such as Kimball (2002) or Biermann (this volume), who conceptualize it as a 're-launched' strengthened UNEP.

Konrad von Moltke, on his part, rejects the idea of a world environment organization and suggests focusing the debate on what he sees as a more feasible reform option: 'clustering' the plethora of multilateral environmental agreements. Von Moltke defines a number of possible clusters—'global atmosphere', 'hazardous substances', 'marine environment' or 'extractive resources'—as well as a set of 'joint institutions' that would facilitate the effectiveness of existing regimes within clusters. These would include joint institutions for scientific assessment, monitoring and environmental assessment, transparency and participation, implementation review and dispute settlement. Von Moltke's chapter is as much a result of the current reform debate as a major contribution to it. It would be interesting to examine in future research whether the creation of clusters will indeed succeed as an alternative to the creation of a world environment organization, or whether the one would rather require the other.

This book neither covers all reform proposals nor does it include all criticisms that have been raised. Drawing on the contributions to this volume as well as other proposals, it seems to us that the debate could benefit in the future from a stronger focus on at least three questions: improved clarity on the conceptual basis for the debate, including its theoretical underpinnings; improved clarity regarding the delineation of the issue area to be addressed; and a better understanding of how smaller organizational and institutional reform proposals intertwine with 'the larger picture'.

Organization, Institution, Regime: Towards Conceptual Clarity

First, we believe that more clarity on the conceptual foundations of the debate on a world environmental organization needs to be reached, in particular on the notion of international organizations, regimes, and institutions. For example, while Oberthür/Gehring and Najam reach convergent conclusions about why organizational reform is unlikely to solve recurrent problems of global environmental governance, they hold opposing views on the necessity to analytically distinguish between regimes and institutions, and organizations.

While Oberthür and Gehring (this volume) view the distinction as 'fiction rather than reality' and hold that regimes and organizations hardly differ in terms of their governing capacity, Najam (this volume) emphasizes the critical importance of the very distinction between institutions and organizations with regard to assumptions about a prospective WEO. In fact, he blames scholars who argue in favour of a new *organization* to trivialize global environmental governance by confusing institutions and organizations. Clearly, this divergence reflects one of the sensitive issues in the international relations literature, which can often be traced to the indifferent usage of these terms—institutions, regimes or organizations—within the social sciences. This may also indicate that the distinction, crucial as it may be, is not as well established as Najam claims.

From a theoretical perspective, part of the WEO debate follows the cleavage within political science between rational institutionalism and sociological institutionalism: while the former largely understands organizations as *institutions* that actors set up to gain from cooperation (which explains why Oberthür and Gehring see no major functional difference between international organizations and regimes), the latter focuses on organizations as *political actors* in their own right that provide information, lobby governments, provide capacity-building programmes, pursue political goals, et cetera.

These different theoretical approaches result in different policy recommendations. One example is Gehring and Oberthür's (2000 and this volume) claim that creation and support of international environmental regimes would be a better way of dealing with the global environmental crisis than creating a new agency. In contrast, many supporters of a WEO would argue that supporting regimes and creating a new agency are not conflicting propositions. A world environment organization could function within a net of international regimes; it would not abolish environmental treaty-making for the same reason that the ILO did not abolish treaty-making on labour rights or the WTO on trade. The question is how these treaties can best be coordinated and how international organizations might assist governments in initiating, negotiating and implementing these agreements.

Most contributions to this volume illustrate this lack of uniform use of concepts and terminology, which necessitates, we believe, more theoretical research and conceptual debate on these issues within the global governance discourse (see Biermann and Bauer, forthcoming, in more detail).

Delineation of Issue Coverage

Second, a major bone of contention is the thematic focus of the debate and reform proposals. Critics of a world environment organization have maintained, for example, that environmental protection is too complex to be addressed by one single agency. Proponents would respond here that almost all countries have established a distinct ministry for the environment, which might indicate that environmental policy can be dealt with by one focal point within an administrative system (see also Biermann, this volume).

A more central issue is the relationship between environmental and development goals, especially of the South. Global governance needs to further sustainable development and not only environmental policy, as emphasized by Najam in his chapter in this volume. Environmental protection cannot be perceived in isolation, and when political agreements on protection of tropical forests or regulation of fossil fuel consumption are negotiated, important questions of economic development are undoubtedly at stake. A world environment organization would need to take this into account. Although its objective would not be to bring about economic development per se, a new organization would have to strive not to impede economic development and to make both policy goals mutually supportive.

Some experts and politicians aim thus at a more wide-ranging integration, particularly the 'merger' of UNEP and UNDP to bring environmental and development concerns in harmony.[1] Such a merger, however, seems unrealistic. In the past, industrialized countries have opposed proposals for an international organization for development policy, and it seems unlikely that they would consent to upgrading UNDP and UNEP into a 'world organization on sustainable development'. As for developing countries, a merger of UNDP and UNEP could be seen as problematic unless they were guaranteed that no resources would be shifted from the development side (that is, UNDP's current budget) to the environmental side (that is, UNEP's current budget). Thus Gustave Speth, a former UNDP administrator, while supporting the creation of a world environment organization, added the caveat that this new organization should by no means assume operational functions in the field, which should remain with the existing bodies, including UNDP (Speth 1998).

Overall, it appears that the exact thematic focus and delineation of governance areas, in particular the institutional and organizational relationship between environmental protection and sustainable development, requires additional research and a fresh focus in the debate on UN reform.

---

[1]   See for example the speech in the German Bundestag delivered by the Social Democratic Party's environmental policy spokesperson on 25 January 1999 (noted in *epd-Entwicklungspolitik* 5/99).

Small Reforms Versus Big Issues

Third, we believe that much energy is currently lost in debates on global governance reform through arguing more about the focus of the debate rather than about problems and solutions. In the case of a world environment organization, this relates in particular to the question of whether such an agency would be 'organizational tinkering' and hence result in neglecting 'the bigger issues'. Clearly, a world environment organization will be no panacea for all problems. The crucial issues are lack of capacity for environmental policy in the developing world and sluggish implementation of obligations by industrialized countries, as pointed out by Najam in his contribution to this book. And as Calestous Juma, the former head of the secretariat to the biodiversity convention, observes, 'There is no guarantee that the new agency will perform better in this regard' (Juma 2000). Yet it is unclear whether the current state of affairs does offer much hope for improvements either. Judged against the weak UNEP, a world environment organization might be in a better position to embark on a new global capacity-building and technology-transfer initiative for the South, and it could be in a better position to influence Northern environmental policies, too.

At the other side of the spectrum, some authors appear to object to proposals for a world environment organization because they see these as reaching too far and to be unfeasible. Yet new international organizations have frequently been created in the last decades, and programmes and even convention secretariats have been upgraded to full-fledged international organizations with enlarged mandate and resources. For example, the United Nations Industrial Development Organization, founded in 1966, was transformed into a specialized agency in 1985. The United International Bureaux for the Protection of Intellectual Property from 1893 (then with a staff of seven) was upgraded in 1974 to the World Intellectual Property Organization as a UN specialized agency, with a staff of now 859 and a comprehensive mandate to administer intellectual property matters of all UN members. The World Trade Organization came into being in 1995 after four decades of highly contentious intergovernmental negotiations under the General Agreement on Tariffs and Trade. Likewise, the International

Criminal Court was established in 2002 despite stiff resistance from powerful governments.

All this suggests that governments are still willing to strengthen international cooperation by setting up intergovernmental bodies. Independent of the desirability of yet another world organization, these examples—together with the support that some reform proposals have mustered among policy-makers—make the eventual launch of a world environment organization not entirely unrealistic. However, important objections have been raised, and supporters of a new agency have work to do to address these objections in a comprehensive and convincing manner.

Outlook

The need to conduct more research as elaborated above relates to the general lack of knowledge on the role and effectiveness of intergovernmental organizations in the social sciences.[2] Both proponents and opponents of a world environment organization have had to build their arguments in most cases on the basis of own personal experiences, theoretical deliberation and general visions, rather than on the findings of empirically-based research programmes. The social sciences have so far largely neglected the study of intergovernmental organizations and the effects they have in world politics. Questions such as the creation and effectiveness of international regimes have received significantly more attention.[3] While there exists some literature of mainly descriptive studies that are informed by international law, diplomatic history and the accounts of practitioners, few academics have yet attempted to address international organizations as actors in their own right.[4] This holds in particular for international environmental cooperation.

---

[2] The following paragraphs are elaborated in more detail in Biermann and Bauer (2004 and forthcoming).

[3] See Haas, Keohane and Levy (1993), Young (1994), Victor, Raustiala and Skolnikoff (1998), Young, Levy and Osherenko (1999), Miles et al. (2002) for but a few examples.

[4] For notable exceptions, see Barnett and Finnemore (1999), Reinalda and Verbeek (1998), Bartlett, Kurian and Malik (1995), Jönsson (1986), Ness and Brechin (1988), Haas (1990), Haas and Haas (1995) and Malik (1995).

We see this neglect of international organizations—as opposed to institutions—in most of the social sciences, notably the study of international relations, as a major flaw. The institutionalist research programme has advanced our understanding of international environmental cooperation and broadened our empirical knowledge through numerous case studies. Yet it has done little to further our understanding of the role that intergovernmental environmental organizations play. Likewise, international lawyers have offered extensive surveys and analyses of the legal-institutional design, diplomatic history and mandates of international organizations, but they have failed to address the effects these organizations have in the real world. Management studies and organizational theories have brought forth a vast literature on institutional dynamics and learning processes. While their findings are commonly used to analyze the successes or failures of both private businesses and non-profit organizations, these have hardly ever been applied to international (environmental) organizations. In short, there is a considerable lack of attention to the role of intergovernmental organizations, such as UN agencies or convention secretariats, in academic analyses of global (environmental) governance.[5]

This persistent neglect of intergovernmental organizations is problematic for at least two reasons. First, the limited understanding of intergovernmental organizations and their effects on international environmental politics is likely to result in misleading conclusions about the state of global environmental governance, with an anachronistic emphasis on sovereign states that encompasses a perception of international organizations as mere passive structures established by states. Consequently, little attention is being paid to the role of intergovernmental organizations as actors with some degree of independence.[6] Second, ignoring the effects international organizations have in international

---

[5] An ongoing multidisciplinary effort in this respect is currently being undertaken by the Global Governance Project's (www.glogov.org) MANUS research group on the effectiveness and learning of intergovernmental organizations; e.g. Bauer (2004), Biermann and Bauer (2004 and forthcoming), Siebenhüner (2003).

[6] For an exception, cf. Reinalda and Verbeek (1998).

environmental politics leaves out a key element of the very structure of global environmental governance. The 'effectiveness' of the United Nations and its specialized agencies in the field of environment and development has been subject to intense *public* debate—with limited systematic academic response, especially in mainstream international relations research. Hence, the very debate about the pros and cons of creating a world environment organization could benefit greatly from a more solid grounding in academia regarding the role of intergovernmental organizations.

It is with this need in mind that we have turned to editing this volume. This collection of eight contributions to the reform debate cannot substitute for a larger empirical research programme on international environmental organizations that we miss in social sciences and international law. Rather, by providing insightful contributions from various angles of the debate over a world environment organization, it aims to illustrate the need for, and indeed to inspire, further research on the organizational aspects of international environmental governance.

References

Barnett, M.N. and M. Finnemore (1999), 'The Politics, Power and Pathologies of International Organizations', *International Organization,* vol. 53 (4), 699–732.

Bartlett, R.V., P.A. Kurian and M. Malik (eds) (1995), *International Organizations and Environmental Policy,* Greenwood Press, Westport.

Bauer, S. (2004), *Does Bureaucracy Really Matter? The Politics of Intergovernmental Treaty Secretariats,* paper presented at the 2004 Annual Convention of the International Studies Association, Montréal, March 2004.

Biermann, F. (2000), 'The Case for a World Environment Organization', *Environment,* vol. 20 (9), 22–31.

Biermann, F. (2002), 'Strengthening Green Global Governance in a Disparate World Society. Would a World Environment Organisation Benefit the South?', *International Environmental Agreements: Politics, Law and Economics,* vol. 2, 297–315.

Biermann, F. (2005), 'The Rationale for a World Environment Organization', in F. Biermann and S. Bauer (eds), *A World Environment Organization. Solution or Threat for Effective International Environmental Governance?,* Ashgate, Aldershot, 117–143.

Biermann, F. and S. Bauer (2004), 'Assessing the Effectiveness of Intergovernmental Organizations in International Environmental Politics', *Global Environmental Change. Human and Policy Dimensions*, vol. 14 (2), 189–193.

Biermann, F. and S. Bauer (forthcoming), *Managers of Global Governance. Assessing and Explaining the Effectiveness of Intergovernmental Organisations*, Global Governance Working Paper, The Global Governance Project, Amsterdam, Berlin, Potsdam and Oldenburg [available at www.glogov.org].

Charnovitz, S. (2005), 'Toward a World Environment Organization: Reflections upon a Vital Debate', in F. Biermann and S. Bauer (eds), *A World Environment Organization. Solution or Threat for Effective International Environmental Governance?*, Ashgate, Aldershot, 87–115.

Elliott, L. (2005), 'The United Nations' Record on Environmental Governance: An Assessment', in F. Biermann and S. Bauer (eds), *A World Environment Organization. Solution or Threat for Effective International Environmental Governance?*, Ashgate, Aldershot, 27–56.

Gehring, T. and S. Oberthür (2000), 'Was bringt eine Weltumweltorganisation? Kooperations-theoretische Anmerkungen zur institutionellen Neuordnung der internationalen Umwelt-politik', *Zeitschrift für Internationale Beziehungen*, vol. 7 (1), 185–211.

Gupta, J. (2005), 'Global Environmental Governance: Challenges for the South from a Theoretical Perspective', in F. Biermann and S. Bauer (eds), *A World Environment Organization. Solution or Threat for Effective International Environmental Govern-ance?*, Ashgate, Aldershot, 57–83.

Haas, E.B. (1990), *When Knowledge is Power: Three Models of Change in International Organizations*, University of California Press, Berkeley.

Haas, P.M. and E.B. Haas (1995), 'Learning to Learn: Improving International Governance', *Global Governance*, vol. 1, 255–85.

Haas, P.M., R.O. Keohane and M.A. Levy (eds) (1993), *Institutions for the Earth: Sources of Effective International Environmental Protection*, MIT Press, Cambridge (Mass.).

Jönsson, C. (1986), 'Interorganization Theory and International Organization', *International Studies Quarterly*, vol. 30 (1), 39–57.

Juma, C. (2000), 'Stunting Green Progress', *Financial Times*, 6 July.

Kimball, L.A. (2002), 'The Debate Over a World/Global Environment Organisation: A First Step Toward Improved International Institutional Arrangements for Environment and Development', in D. Brack and J. Hyvarinen (eds), *Global Environmental Institutions. Perspectives on Reform*, Royal Institute of International Affairs, London, 19–31.

Kirton, J. (2005), 'Generating Effective Global Environmental Governance: The North's Need for a World Environment Organization', in F. Biermann and S. Bauer (eds), *A World*

Environment Organization. Solution or Threat for Effective International Environmental Governance?, Ashgate, Aldershot, 145–172.

Malik, M. (1995), 'Do We Need a New Theory of International Organizations?', in R.V. Bartlett, P.A. Kurian and M. Malik (eds), *International Organizations and Environmental Policy*, Greenwood Press, Westport, 223–37.

Miles, E., A. Underdal, S. Andresen, J. Wettestad, J.B. Skjærseth and E. Carlin (2002), *Environmental Regime Effectiveness. Confronting Theory With Evidence*, MIT Press, Cambridge (Mass.).

Najam, A. (2005), 'Neither Necessary, Nor Sufficient: Why Organizational Tinkering Will Not Improve Environmental Governance', in F. Biermann and S. Bauer (eds), *A World Environment Organization. Solution or Threat for Effective International Environmental Governance?*, Ashgate, Aldershot, 235–256.

Ness, G.D. and S. Brechin (1988), 'Bridging the Gap: International Organizations as Organizations', *International Organization*, vol. 42 (2), 245–73.

Newell, P. (2001), 'New Environmental Architectures and the Search for the Effectiveness', *Global Environmental Politics*, vol. 1 (1), 35–44.

Newell, P. (2002), 'A World Environment Organization. The Wrong Solution to the Wrong Problem', *The World Economy*, vol. 25 (5), 659–71.

Oberthür, S. and T. Gehring (2005), 'Reforming International Environmental Governance: An Institutional Perspective on Proposals for a World Environment Organization', in F. Biermann and S. Bauer (eds), *A World Environment Organization. Solution or Threat for Effective International Environmental Governance?*, Ashgate, Aldershot, 205–234.

Reinalda, B. and B. Verbeek (eds) (1998), *Autonomous Policy-Making by International Organizations*, Routledge, London.

Siebenhüner, B. (2003), *International Organisations as Learning Agents in the Emerging System of Global Governance. A Conceptual Framework*, Global Governance Working Paper no. 8, The Global Governance Project, Amsterdam, Berlin, Potsdam and Oldenburg [available at www.glogov.org].

Simmons, B.A. and L.L. Martin (2002), 'International Organizations and Institutions', in W. Carlsnaes, T. Risse and B. Simmons (eds), *Handbook of International Relations*, Sage, London, 192–211.

Speth, J.G. (1998), Interview with Jens Martens, Bad Honnef, Germany, July 1998 (at http://bicc.uni-bonn.de/sef/publications/news/no4/speth.html).

Victor, D.G., K. Raustiala and E.B. Skolnikoff (1998), *The Implementation and Effectiveness of International Environmental Commitments. Theory and Practice*, MIT Press, Cambridge (Mass.).

von Moltke, K. (2001), 'The Organization of the Impossible', *Global Environmental Politics*, vol. 1 (1), 23–28.

von Moltke, K. (2005), 'Clustering International Environmental Agreements as an Alternative to a World Environment Organization', in F. Biermann and S. Bauer (eds), *A World Environment Organization. Solution or Threat for Effective International Environmental Governance?*, Ashgate, Aldershot, 175–204.

Young, O.R. (1994), *International Governance. Protecting the Environment in a Stateless Society*, Cornell University Press, Ithaca (NY).

Young, O.R., M.A. Levy and G. Osherenko (eds) (1999), *Effectiveness of International Environmental Regimes: Causal Connections and Behavioral Mechanisms*, MIT Press, Cambridge (Mass.).

Index

Printed and bound by CPI Group (UK) Ltd, Croydon, CR0 4YY

21/10/2024

01777084-0008